DESIGNING FOR OLDER ADULTS

Principles and Creative Human Factors Approaches

DESIGNING FOR OLDER ADULTS

Principles and Creative Human Factors Approaches

Arthur D. Fisk, Wendy A. Rogers,
Neil Charness, Sara J. Czaja and
Joseph Sharit

CRC PRESS

Boca Raton   London   New York   Washington, D.C.

Library of Congress Cataloging-in-Publication Data

Designing for older adults : principles and creative human factors
approaches / Arthur D. Fisk ... [et al.].

p. cm.
Includes bibliographical references and index.
ISBN 0-415-28611-5
1. Human engineering. 2. Aged. I. Fisk, Arthur D.

TA166 .D45 2004
620 . 8'2—dc21   2003010735

Visit the CRC Press Web site at www.crcpress.com

© 2004 by CRC Press LLC

No claim to original U.S. Government works
International Standard Book Number 0-415-28611-5
Library of Congress Card Number 2003010735
Printed in the United States of America 2 3 4 5 6 7 8 9 0
Printed on acid-free paper

We dedicate this book to:

-   The older adults who have inspired us – our parents and grandparents
-   The participants in our research studies whose invaluable efforts have helped us to develop these guidelines

Contents

List of figures

List of tables

Preface

Our goal for this book was to provide a primer on the issues that must be considered when designing systems, products, or environments for older adults. Because our target audience is the design community, we wanted to provide guidelines for design that were accessible to a wide variety of readers and immediately applicable to the design process.

There is quite a large academic literature on age-related changes in abilities, learning, and performance and how such changes relate to interactions with technology. Our challenge was to translate this information into guidelines and to present the key information in a readable format. Our approach to writing this book was to step outside our traditional style of writing academic books and research articles. The book does not contain a lot of specific references to support each point or guideline that we provide. Instead, we present our collective views based on many years of research in the field of aging.

It is somewhat unusual to have a book coauthored by five people. Rather than develop an edited volume in which we each write chapters on different topics, we worked together on the entire content of this book. As a team, we developed the structure of the book, its main ideas, and the guidelines for each topic. The information we present is intended to be specific enough to be immediately applicable, yet general enough to be relevant to technologies of the future that cannot even be imagined. Although future cohorts of older adults may have more experience with some technologies, general age-related changes in capabilities will still occur and, given the dynamic nature of technology, there will always be the need for older adults to learn to use new technologies.

We firmly believe that successful design for older adults will result from an understanding of many things, including the typical changes that accompany aging, how awareness of such changes should influence the design process, and a human-factors approach to development and testing of products, systems, and environments. For ease of exposition, we use the term *human factors* and assume it to be synonymous with *ergonomics*. Other related terms that denote similar perspectives are *engineering psychology, applied experimental psychology*, and *industrial engineering*. The discipline of human factors is a multidisciplinary approach to design that puts the user at the center of the design process with the goal of developing safe, effective, and efficient user-system interactions. Our team is also multidisciplinary, being composed of two industrial engineers and three psychologists.

This book represents the combined efforts of the principle investigators of the Center for Research and Education on Aging and Technology Enhancement (CREATE). CREATE is sponsored by the U.S. National Institutes of Health (National Institute on Aging) through Grant P01 AG17211. CREATE is directed by Sara Czaja at the University of Miami. The other CREATE investigators are Joe Sharit, also at the University of Miami; Neil Charness at Florida State University; and Wendy Rogers and Arthur (Dan) Fisk at the Georgia Institute of Technology.

CREATE is a multidisciplinary, collaborative center dedicated to solving the problems of aging and technology use. The development of the Center was motivated by the increased number of older people in the population and the increased reliance on technology in most societal contexts. It is estimated that people beyond the age of 65 will represent 22% of the population by 2030. The overarching goal of CREATE is to help to ensure that current and future generations of older adults will be able to use technology successfully and that the potential benefits of technology can be realized for older populations. To that end, this book represents the development of comprehensive design guidelines for the design of existing and emerging technologies that may be used by older adults.

The Scientific Advisory Board of CREATE consists of Jim Baker, Colin Drury, Jim Fozard, Melissa Hardy, Bill Howell, Beth Meyer, John Thomas, and Richard Schulz. We appreciate their guidance, advice, and support during the development of this book. In addition to support from the NIH (NIA), we acknowledge support from the U.S. National Institute for Occupational Safety and Health and, in particular, the guidance of James Grosch. We also appreciate the support we receive from our academic homes: the Georgia Institute of Technology, Florida State University, and the University of Miami.

We extend our deepest appreciation to all the researchers, graduate students, and postdoctoral fellows who have been involved in CREATE. Although there are too many people to list by name, we want them all to know how much we value our collaborations with them and how much they have contributed to the development of this book.

The CREATE Team
Arthur D. Fisk
Wendy A. Rogers
Neil Charness
Sara J. Czaja
Joseph Sharit

Part I
Fundamentals

Chapter 1

Toward better design for older adults

Why design for older adults? There are many reasons. Clearly more and more consumers and users of technology are joining the ranks of "older adult." Indeed, to emphasize the fact that life expectancy is dramatically increasing, Rowe and Kahn estimated in their 1998 book that of all the people who have ever lived to be 65 years of age or older, more than half are alive today. Such a change in demographics brings with it important changes in the demands for products and services. Addressing those demands is not just the role of technology but, more important, we would argue, is the role of those who determine how that technology should function. That is, designers hold the key, in many instances, not only to increasing market share for a given product but to increasing the quality of older adults' lives.

This text provides a practical introduction to human factors and older adults. The book is aimed toward professionals working to develop systems and environments better to accommodate the needs of older adults. It should also be useful to people interested in the design process as it relates to older adults. To that end, we emphasize the application of the scientific knowledge base concerning age-related issues of perception, cognition, and movement control. The text is meant as a reference source, with practical guidelines and advice for design issues ranging from lighting, computer input device selection, and Web site design to training program development, work task design, and health care technology development.

There are many research-oriented publications available. What we found missing was a reference document, grounded in the current state of scientific knowledge, accessible to the broad audience of product designers, health care practitioners, managers, and others who need answers derived from the scientific knowledge base but translated for more immediate applicability. Our motivation for developing this reference guide was to ensure that optimal recommendations for design were provided in an accessible format. An important caveat is that the recommendations are based on the current state of knowledge.

As mentioned, society is "getting older." As can be seen in Figure 1.1, this aging of society is occurring worldwide. Within Asia, North America, and Europe, the current percentage of the population older than age 65 ranges from 6 to 15 percent. It is estimated that by 2030, those figures will range from 12 to 24 percent. The fastest growing subgroup represents those beyond 80 years of age. People are living longer, remaining more active into older age, and remaining in their homes longer before finding the need for "assisted living" arrangements. Aging brings with it changes in perception, cognition, and control of movements. We address these changes as they relate to design in the following chapters of the book.

In addition to changes in demographics, there has been an enormous change in technological capability. There have been changes in the way products work, look, act, and react to people who use them. These changes in technology, coupled with changing capabilities of the people using the technology, can lead to less-than-desirable interactions with products. We have conducted numerous focus group and survey research studies and have found that the range of technology encountered in the daily lives of older adults is very broad. Unfortunately, the extent of frustration encountered in their dealing with this technology is also quite evident. Proper attention to design will eliminate much of this frustration.

Our research has also made it clear that although older adults do have unique usability constraints as compared to those of younger adults, these usability problems are often shared among age groups. When usability is improved for older adults, it is also improved for younger adults. What is also clear from human-factors research is that improved usability will enhance market penetration of a given product. Improved usability will improve quality of life and, with some classes of products, can save lives.

What products do older adults use?

Age does not necessarily limit the number of products used. One of our focus group studies demonstrated the fallacy to the myth that older adults wish to avoid new technology. It has been reported that

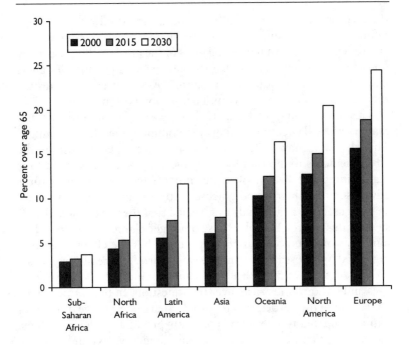

Figure 1.1   Percentage of the population older than age 65, currently and
estimated for the future. (Source: Kinsella and Velkoff, 2001).

older adults are less likely than younger adults to use technology.
This usage difference may be mediated by age-related income
disparities, perception of actual need to use the technology, products
being too difficult to learn to use, and other factors. Our research
has shown that when there is a need to use technology, older adults
wish to use that technology; however, the benefits must be made
clear.

New technology notwithstanding, what can we say about usage
of products in general? How does age affect usage of common
everyday products? In further research, we asked people of all ages
about the frequency with which they used products across a variety
of everyday activities. We found that individuals of all age groups
used a large percentage of products across various categories (e.g.,
tools, cleaners, over-the-counter medications, toiletries, health care
products).

Products are used by adults of all ages, but are those products
easy to use? It is commonly believed that many household products

that we use are "user-friendly." Product instructions and usage information on common everyday products seem easy to comprehend and remember, at least to the designers of the product. Because many household products are familiar to many people, product designers may assume that product use is simple. As a consequence, testing their actual usability may be minimal or nonexistent.

This thinking may lead to an under-appreciation of the difficulties that people will encounter using even common household products. This thinking may even transcend product types. Consider the advertisements found on blood glucose meters, devices critical for maintaining the health of some individuals. One manufacturer advertised that their blood glucose meter required only three easy steps to use. Yet, a task analysis revealed that there were actually 52 steps in this process! Similar "ease of usage" claims can be found on hundreds of common household products but the claims are often not based on assessments of how easy the products really are to use. We conducted a study with a large sample of individuals of different ages and backgrounds to answer that question. In the study, almost 75 percent of the participants reported experiencing usage difficulties while using many everyday, common products. User problems included difficulties in seeing or comprehending text and symbols on the product, problems in remembering instructions or warnings, or movement-control difficulties related to holding or opening the product.

Age may exacerbate usability problems or, in the very least, increase the consequences of problems. For example, falling may be an inconvenience for a younger adult, but it can be a life-threatening event for an older adult.

Is good design worth the effort?

In an effort to understand better whether attention to characteristics of design to improve usability can improve the lives of older adults, we should return to the series of focus groups we conducted, which documented the usability problems encountered in the daily activities of older individuals. The difficulties that participants reported were classified according to the activity engaged in when the problem was encountered; the source of the problem (i.e., motor, visual, auditory, cognitive, external, or general health limitations); whether the problems were related to the inherent difficulty of tasks or potential negative outcomes; and how participants responded to a certain problem (e.g., stopped performing the task in response to their limitations, or compensated somehow).

Of the problems reported by the older adults, 47 percent were due to financial limitations, health difficulties, or other general concerns. Each remaining problem was classified according to whether it could potentially be solved through redesign, training, or some combination of the two. Approximately 25 percent of the problems could eventually be remedied by improving the design of the systems. Such redesign efforts could be applied to solve sensory or motor problems and might involve changes, such as lowering steps on buses, developing tools for grasping or scrubbing, improving chair design, or enlarging letter size on a label.

The remaining 28 percent of the reported problems had the potential to be solved through the provision of training or through a combination of training and redesign. For example, an older person learning to drive for the first time would benefit from driver training (redesign of the automobile would not be necessary). However, for more complex systems, such as the use of a personal computer or health care technologies, training would be needed for a novice older adult, and there is clearly the potential for design improvements that would increase usability of such systems for users of all ages. These data seem quite striking to us: more than 50 percent of the problems of daily living reported by that sample of older adults were potentially addressable through human-factors design efforts. We may find that the 47 percent of the problems that do not appear initially to be directly addressable through human-factors interventions could be lessened through proper application and design of technology. Consider, for example, the promise that properly designed technology has for aiding some health-related problems or problems due to older adults being far away from family and friends.

Defining "older adult"

If forced to "give a number" to the question of when a person is an "older adult," we would most likely say that older adults are those individuals who are 60 years of age and older. However, it is important to keep in mind that such classification is not always so straightforward. There are no definitive boundaries between what is considered young and what is considered old; thus, age is not easily represented as a nominal variable. As a result of reporting differences across databases, some of the graphs we include in this book represent older adults as older than age 50 or older than age 65.

Chronological age itself is useful only as a marker for corresponding behavioral changes. As a scientific understanding of behavior can best be achieved through a careful analysis and description of change, a major goal of current aging research is to identify specific changes that occur throughout the aging process. For example, the detection threshold for the normal range of human speech goes through a more rapid decline after age 60. In vision, accommodation is severely limited by the age of 65, resulting in difficulty in tracking objects at varying distances, but visual acuity for reading small print declines for many people by age 40.

Aging occurs on many levels and can be categorized on at least three dimensions: biological, psychological, and social. Simple observation should make it clear that older adults, however defined, are not a homogeneous group: individual differences are prevalent regardless of the decade of adult life. Until a much more complete understanding of age-related changes is available, researchers and practitioners generally rely on chronological age as an index. We believe that it is useful to think of two groupings of "older adults." One grouping is what we term the *younger-old*, ranging in age from approximately 60 to 75. The other grouping we call the *older-old*, comprising those individuals beyond 75 years. Of course, adding to the difficulty of defining older adult is the fact that we must consider the task or situation: what is young for some activities may be old for others.

Why then do we think about older adults and not just individuals? Generally, older adults do have much in common in terms of the biological, psychological, and social dimensions, even though they do demonstrate individual differences. When we are considering design, we are focusing on those similarities that allow us to optimize the design. We must also be aware of the individual differences so that we can determine whom our design can and cannot accommodate.

Toward better design

The objectives of this book are centered on improved design, especially for products, services, and environments used and encountered by older adults. The primary audience for the book is those individuals involved in making design decisions. These decisions might be associated with Web pages, work tasks, training programs, health care technologies, and so on. We believe the book will also be helpful to those who make decisions concerning living environments, such as lighting or navigational signage.

Our goal is to translate scientific knowledge into information that may be developed into "best practices." However, the science is better developed in some areas than in others. Therefore, the guidelines are more explicit and detailed in some chapters than in others. Moreover, it is always important to consider the task and the context in which a given recommendation will be applied. For example, when we provide recommendations regarding assisting the aging eye, we would recommend 12-point font size for labels on products. However, if the design task involves Web pages, 12-point font may slow down reading and increase the need for scrolling so a trade-off might be 10-point font in such a usage context. Additionally, if the text will have to be read from a distance, font sizes larger than 12-point would be recommended.

General guidelines, specific questions

In most instances, the guidelines we supply provide a starting point for good design. The information is certainly necessary for good design but should never be thought of as sufficient without final user testing. As an analogy, anthropometry data might give someone a starting point for determining proper shelf height but, for each situation, one would still be required to perform "fitting trials" to ensure proper height for the users of the shelving. In Chapter 3, we discuss and provide details concerning basic issues in design and usability testing. This chapter is meant to be a primer concerning the tools and techniques important for analyzing the prospective users' needs, the possible system capabilities, and testing to determine whether the needs and capabilities of such persons actually match the proposed system, environment, or training. Designers should be familiar with this material before attempting design for older adults.

The material in this book is a compilation of our knowledge, research, and experience. Use of this information cannot be guaranteed to fit every design problem encountered. We focus on application of the scientific knowledge base in a general sense.

Overview of the book

Although no book can serve as a sole source representing all the subfields of human factors and aging, we have attempted to provide a broad coverage of important usability topics. The book is divided into three main sections: Fundamentals (Chapters 1–3), Design Guidelines (Chapters 4–7) and Exemplar Applications (Chapters 8–9).

In Fundamentals, Chapter 1 provides the background and purpose of the book. Chapter 2 provides more details concerning characteristics of older adults. In that chapter, we briefly review psychological characteristics that are important to consider during the design process. Chapter 3 is the primer on human factors tools and techniques.

The Design Guidelines' chapters each target a specific domain, providing an overview of key findings related to aging in that domain as well as specific design guidelines. Chapter 4 provides practical translations of scientific data concerning perception into design recommendations. Chapter 5 is concerned with the development and implementation of training and instruction. The chapter provides material relevant to creation of workplace training as well as instruction for use of technology. Chapter 6 offers practical translations of scientific data concerning input and output devices into recommendations for design and for the selection of input and output devices. Chapter 7 is concerned with the design of system interfaces, including issues relevant to human–computer interaction (e.g., personal computers, the Internet) as well as facilitating interaction with systems, such as automated teller machines, library systems, microwave panels, cell phone menus, and hand-held devices.

The two Exemplar Applications chapters provide examples of how the design guidelines are applicable in two broad areas: work and health care. Chapter 8 addresses making the work environment age-friendly and concerns issues of structuring work tasks for an aging workforce, of information flow, of work pacing, and so on. We address health care technology in Chapter 9. Health care is a critical concern to adults of all ages but especially to older adults, given their propensity to illness and chronic conditions. Technology offers the potential to help older adults to be actively involved in the management of their health and health care and to facilitate health care delivery to older adults and their families. However, for such technology to be effective and safe, the attention to human factors is crucial. We chose work and health care as two representative domains, but the design guidelines are applicable in a broad range of domains, such as leisure activities, communication, education, transportation, and home environments.

We conclude the book (Chapter 10) with a brief summary of the themes that emerged throughout all the chapters. These themes illustrate the key issues that must be considered in designing for older adults.

Recommended readings

At the end of each chapter, we provide a short list of articles and books that will supplement the materials discussed in the chapter. For this chapter, we suggest the following additional sources.

Czaja, S.J., Sharit, J., Charness, N., Fisk, A.D., and Rogers, W.A. The Center for Research and Education on Aging and Technology Enhancement (CREATE): A program to enhance technology for older adults. *Gerontechnology* 1 (2001), 50–9.

Fisk, A.D. Human factors and the older adult. *Ergonomics in Design* 7, 1 (1999), 8–13.

Fisk, A.D., and Rogers, W.A. *Handbook of Human Factors and the Older Adult*. Orlando: Academic Press, 1997.

Rogers, W.A., Meyer, B., Walker, N., and Fisk, A.D. Functional limitations to daily living tasks in the aged: A focus group analysis. *Human Factors* 40 (1998), 111–25.

Rogers, W.A., and Fisk, A.D. Human factors, applied cognition, and aging. In F.I.M. Craik and T.A. Salthouse (eds), *The Handbook of Aging and Cognition*. 2nd edn. Mahwah, NJ: Erlbaum, 2000:559–91.

Rowe, J.W., and Kahn, R.L. *Successful Aging*. New York: Pantheon, 1998.

Chapter 2

Characteristics of older adult users

What is the concept of human factors? How can an understanding of the science of human factors and engineering psychology aid people in designing products and aid in evaluating issues relevant to proper design? We briefly answer these questions in this chapter and then address basic issues in perception, cognition, and control of movements that are important to consider in designing products, environments, systems, and training.

What are the underpinnings of human factors?

Too often, we hear people argue that they are human and thus know all they need to know about human factors. They believe that issues addressed by human-factors specialists can be solved by simple common sense. Of course, some issues may be related to common sense, but often common sense is not sufficient to understand details of human behavior relevant to design. Moreover, common sense beliefs may differ across individuals as a result of their experience, education, and culture. An understanding of perception, cognition, and movement control is critical for the human side of the design process.

The background and underpinnings of the field of human factors and engineering psychology illustrate the relevance of this scientific field to the design process. The discipline of human factors can be defined as the study of the characteristics of people and their inter-actions with products, environments, and equipment in performing tasks and activities. The basic tenet of the discipline is that human characteristics must be considered in designing and arranging systems and devices that humans use. The field of human factors develops the scientific knowledge base concerning the capabilities and limitations of people and then uses that scientific knowledge about human behavior in specifying the design and use of a human-machine (or human-environment) system. We may define the overarching goal of human factors as making human interaction with systems and environments one that reduces error, increases productivity, promotes safety, and enhances comfort. The ultimate goal of the science and the practice of human factors is to ensure that human–system and human–environment interactions will be safe, efficient, and effective.

What human characteristics must be considered?

Some reflection on one's own interaction with products, with instructions, with work-related tasks, and so on should lead to some sense of the number of movement-control, perceptual, and cognitive factors recruited when individuals interact with products. Certainly visual and auditory capability is often a crucial factor. Understanding movement capabilities and cognition are also critical to facilitating proper human-centered design. Indeed, when serious accidents related to products are considered, a majority are found to be due to informational causes (processing the perceptual cues, understanding that perceptual information, and responding to it appropriately).

Sensation is the awareness of simple properties of stimuli, such as color. Perception is the awareness of more complex characteristics of the stimuli. Seeing the color red would be sensation, but seeing and recognizing an apple is perception. *Perception* is used here to refer to the activation of the sensation cells, such as the retinal cells, and to the interpretation of that information by calling on stored memory. In this latter sense, the concept of cognition and perception overlap. However, sensation and perception are the first of many complex processes that occur when an individual initiates a behavior. No perception gives complete, direct knowledge of the outside world; rather, cognition takes the products of perception and provides interpretation. *Cognition* refers to all the processes by which the brain takes sensory input, whether from the eyes, the ears, and the like, and transforms, reduces, elaborates, stores, recovers, and uses that sensory input. *Movement control* is a broad term that describes physical responses, such as turning a knob, pressing a lever, or selecting keys with different fingers.

The human-factors approach involves using scientific knowledge about people's capabilities and limitations to create designs capitalizing on strengths and capabilities while guarding against limitations. Age brings with it many capabilities, such as increased wisdom, experience, and knowledge. However, limitations that are associated with perception, cognition, and the control of movements increase in prevalence as one ages. It is important to become aware of these limitations.

The focus of the remainder of this chapter is a review of age-related changes that occur in sensing and perceiving information, processing that information, and physically responding to the information (Table 2.1). This review is not exhaustive, and we provide citations to additional reference material that should be consulted.

Aging and the sensory modalities

Sensory processes have received considerable attention in the investigation of the effects of age on capabilities associated with various activities. We focus only briefly on taste and smell. The remaining three senses customarily studied—kinesthesis, audition, and vision—are discussed in more detail. These senses are more pertinent because they may represent user capabilities and limitations that are directly relevant to design. Auditory and visual capabilities as they relate to design issues are also addressed in depth in Chapter 4.

Taste and smell

Taste and smell are not addressed beyond noting that they show age-related decrements that frequently result in an inability among the very old to distinguish among various foods or odors. In the case of taste, the tongue loses taste buds with age. The evidence indicates that until age 60, the ability to perceive sweet, sour, bitter, and salt—the four basic tastes—does not change much at all. During the 60s, some gradual diminishment is noticed, most particularly with salty flavors. Other things, most notably changes in the sense of smell, often cause in older adults the inability to distinguish various tastes or flavors. Anyone who has had a cold knows that food may be tasteless when the nose is stuffy and the sense of smell is impaired.

Research on age-related declines in the sense of smell is contra-dictory, but most investigators contend that there is some minor

Table 2.1 Description of categories discussed in Chapter 2

Term	Definition	Examples
Sensation, perception	Sensation is the awareness of simple properties of stimuli such as color; it refers to the activation of sensation cells (e.g., retinal cells). Perception is the awareness of complex characteristics of things in the environment; it refers to the interpretation of information that results from sensation	Sensation is seeing the color red or hearing a high-pitched sound. Perception is recognizing the red object as an apple or determining that the sound is an alarm
Cognition	Processes by which the brain takes sensory information from the ears, eyes, etc., and transforms, reduces, stores, recovers, and uses that information	Thinking, problem solving, reasoning, and decision making
Movement control	The carrying out of an action on the basis of perception or cognition; requires the coordination of muscles for control of motion of some type	Steering a car; "double-clicking" a mouse button; grabbing an object from a shelf

deterioration over time. Throughout life, odor recognition varies greatly among individuals. Some older people never had much of a sense of smell; some experience a decline in their later years, whereas others do not seem to perceive a change. When a decline does occur, the root cause may be an atrophy of olfactory bulbs in the nose.

Kinesthetic sensitivity

Why do older people seem to fall often, or why do they sometimes appear to be less stable in movements as compared to younger adults? The answers are at least in part linked to changes in somatic (or kinesthetic) sensitivity. Over time, as we age, the automatic integration of movement-related sensation mediated by vestibular cues for maintaining balance does seem to deteriorate.

Some examples may better explain kinesthetic sensitivity. Few young people have any difficulty in recognizing when they are sitting upright or partially prone, nor do they often make mistakes when called on to locate their feet relative to their knees, such as when climbing on uneven terrain. Usually, they are able to make generalized postural adjustments when getting to their feet, compensating for slight misalignments without giving the matter any thought. Some older adults, on the other hand, are not able unconsciously to control body position or movement, because the loss of their kinesthetic senses leaves them vulnerable to accidental falls and postural instability. For each of us, the sense of movement, touch, and position depends in part on receptors located in muscles, joints, and the skin. For various reasons, some of which can be traced to sensory impairments and some to a breakdown of the brain's integrative capacities, the dizziness reported by some older adults is often attributed to decrease in functioning in these receptors as well as to the integration of visual cues with the receptor information. For our purposes, it is important to recognize that the sense of movement, touch, and position is more variable across an older adult population as compared with a younger counterpart.

Audition

The ability to hear may affect one's ability to interact successfully with systems and to function safely and effectively in environments. If auditory information is an important aspect of design, age-related changes in audition must be considered. Various estimates suggest approximately 10 percent of all middle-aged adults suffer hearing losses of a magnitude that hinders social interaction. By age 65 or so, the percentage has jumped to more than 50 percent of all men and 30 percent of all women. The differences between men's and women's hearing capabilities may change as more women engage in work and, as younger adults, in leisure activities that are detrimental to hearing.

Normally, young adults can hear pure tones in frequencies up to 15,000 vibrations per second. Beginning around age 25, presbycusis — an age-related loss of the ganglion cells necessary for conduction— causes erosion of the upper threshold and, after age 65 or 70, sounds above 4,000 vibrations per second may be inaudible. Conversely, low-range tones (below 1,000 cycles) do not appear to be appreciably affected by age. Volume, or loudness measured in decibels, is a more common measure of hearing. Humans can hear sounds well below the level of a whisper, which averages approximately 8 decibels, to those in excess of 130 decibels, although pain and nausea are associated with the latter. The range of normal conversation is around 60 decibels, and severe hearing impairments are said to occur when an individual's threshold exceeds 35 decibels.

Age-related declines in the hearing of high-frequency sounds may be traced to a deterioration of receptor hair cells, neurons, and vascular changes in the inner ear or in membranes within the inner ear. Depending on which decline is most prominent, the ability to hear different types of sound will be affected. To summarize: the best evidence at this time supports the contention that the association between age and hearing loss of all types is strong. Designers of products and environments must be cognizant of changes in hearing capability influencing older adults' ability to detect tones and other sounds and of changes in the ability to comprehend speech.

Vision

Visual impairments affect many people, regardless of age. However, like so many other chronic conditions, the prevalence of visual impairments increases with age. In fact, age is the single best predictor of visual limitations or blindness. If we live long enough, nearly all of us will have vision problems.

Visual acuity is the measure most often used when speaking of vision. Often it is reported in terms of the ability to discriminate test objects at a distance of 20 feet. Individuals having 20/20 vision can read an eye chart at 20 feet, which is considered normal at that distance, with larger denominators (e.g., 20/20 versus 20/50) indicating progressively poorer eyesight. For example, vision rated as 20/70 means objects distinguishable at 20 feet with impaired eyesight can be discriminated at 70 feet by the normally sighted. Resulting from changes in the eye, most declines become noticeable by the time affected persons are in their late 40s, if not earlier. Hence, seven of ten people beyond age 45 find it necessary to wear glasses, compared to three of ten younger than 45. Fewer than 20 percent of those older than age 65 cannot obtain a corrected vision of at least 20/40, more than adequate for normal activities. However, even with corrected vision, the requirement for "bifocals" may make certain tasks more demanding for those who wear them than for those individuals who do not require bifocals.

Presbyopia, the inability to change the eye's focal length, is so common during the last half of life that most people older than 40 have experienced it. A similar decline in the eyes' ability to adapt to darkness tends to inhibit reading as well as driving at night among older adults. However, carefully controlled illumination can minimize a large share of the problems that might otherwise interfere with a person's daily routine.

Regardless of a person's ability to read a newspaper, changes in vision also affect depth perception and sensitivity to glare. Adapting to bright illumination is difficult for older adults as compared to younger adults. Declines in acuity and hypersensitivity to glare also lead to decrements in an older person's sense of depth and item relationship. Some deterioration in the size of the visual field has also been observed. Finally, research has indicated an apparent slowing in the speed with which visual information is processed, a change that increases with age. Consequently, perceptual flexibility in visual sensation undergoes a gradual decline with age.

Cognition

Interactions with products may be analyzed in terms of various cognitive processing components required for successful performance. For our purposes, we discuss the cognitive components in the framework of human information processing. Age-related changes in cognition can be important to consider in designing for older adults. As with the review of sensory and perceptual changes, we highlight aspects of cognition that designers should consider. Citations to further reference materials are provided at the end of the chapter. Table 2.2 provides a glossary of the scientific constructs discussed in the section on cognition.

Memory

A common belief is that memory gets worse as we get older. However, whether memory capability declines with age depends on what kind of memory is involved in any given activity. Age-related decline in working memory (sometimes called *short-term memory*) is well documented. *Working memory* refers to the capability to keep information active temporarily while we "work on it" or until we use it. Using a telephone menu system may require extensive working

Table 2.2 Definitions of cognitive constructs

Term	Definition
Attention	The selective process that controls awareness of events in the environment; determines the events to which we become conscious; limited, operating selectively on stimuli in the environment. For example, if a person in the midst of multiple conversations can "pay attention" to only one particular conversation, this has the effect of screening out the other conversations to some extent. What is also apparent, apart from this screening process, is that we are not, generally, able to listen to two conversations at the same time. If someone loudly calls our name (a salient stimulus), our attention is pulled away from the conversation to the source of our name, and we miss what the person said in that original conversation.
Dynamic visual attention	The ability to integrate from a large visual space information that cannot be comprehended in a single glance.
Language comprehension	The ability to interpret verbal information, whether written or spoken; includes the ability to understand individual words, to understand sentences and paragraphs, and to draw logical inferences that are implied in a text or discourse .
Procedural knowledge	Knowledge about *how* to perform activities; varies along the dimension of automaticity, from knowledge that is executed almost without thought (e.g., shifting gears or steering a car) to explicit but well-practiced routines (e.g., following a recipe).
Prospective memory	Remembering to perform an action in the future; time-based prospective memory tasks are those in which a person must remember to do something at a certain time (e.g., at 2:00 PM) or after a particular amount of time has passed (e.g., in 2 hours); event-based prospective memory tasks are those in which something must be done in response to an event (e.g., when the buzzer goes off, turn off the oven).
Semantic memory	Long-term memory for acquired knowledge; includes such concepts as vocabulary, historical facts, cultural norms, rules of language, art and music information, and the like.
Spatial cognition	The ability to manipulate images or patterns mentally, to represent information and transform it (e.g., mentally rotate an image) or to accurately represent spatial relationships among components.
Working memory	Memory of what has just been perceived and what is currently being thought about; consists of new information and information that has recently been "retrieved" from long-term memory. Only a few bits of information can be active in working memory at any one time (think of holding three names in memory versus ten names). Information held in working memory decays quite rapidly unless it is rehearsed to keep it there.

memory if the user of that system needs to remember the first option while five or six other options are announced. Designing to guard against the limits of working memory is an important recommendation.

Working memory capacity affects performance of real-world tasks to varying degrees. Age differences in a variety of domains (speech and language comprehension, reasoning, problem solving) have been attributed to age-related differences in working memory. These conclusions have been reached through both within-context and out-of-context assessments of working memory. Within-context assessments involve inferring working-memory capacity from task performance. For example, in a concept formation task, redundant questions asked by a person might be indicative of a working-memory deficit. An out-of-context assessment of working memory involves measuring performance on tasks specifically designed to assess memory (e.g., span tasks found in a psychological ability test, keeping-track tasks). Regardless of the measurement method, a reduction in working memory is typically found in older adults.

Another kind of memory often is termed *long-term memory*. Long-term memory can be thought of as a more permanent storage of knowledge (including learned movements and skilled behaviors). A type of long-term memory—semantic memory—does not appear to decline with normal aging. *Semantic memory* is defined as the store of factual information that accrues through a lifetime of learning. Remembering the meaning of a word is a form of semantic memory, as is remembering historical facts, memory for art and music, and general knowledge: basically, information acquired throughout one's lifetime. Designing to make use of such semantic memory can be important. Population stereotypes (such as up being "on" for a light switch in North America but "off" in Europe) are a form of semantic memory shared by groups of people. Making use of population stereotypes in design can facilitate ease of use. Design that is contrary to population stereotypes can lead to disastrous consequences.

Another form of long-term memory is called *prospective memory* (i.e., memory to remember to do something in the future). If our prospective memory is based on doing something at a later time (such as remembering to take medication in 4 hours), this is time-based prospective memory. Another kind of prospective memory depends on performing some action after some event occurs (such as remembering to take medication after eating or after a timer bell sounds). This latter kind of prospective memory is called *event-*

based prospective memory. Age-related declines in prospective memory are usually much greater for time-based than for event-based tasks. From a design perspective, it is important to guard against time-based prospective memory demands and to ensure that event-based prospective memory demands are coupled with an event that will provide the appropriate reminder to cue memory.

Visual attention

Many parts of the process of interacting with products involve visual search. Detecting and categorizing warning information or finding which buttons to press on a new automated teller machine are examples of search-detection tasks. Searching for things requires attention. We find that as the demands on attention increase, age-related performance problems also increase.

Dynamic visual attention is another aspect of cognition that can be related to successful interaction with products or environments. Dynamic visual attention is how we scan the environment and involves focusing attention in one location and then another location. However, the ability of a person to focus attention and then reorient that focus is limited by the availability of a finite amount of attention-related resources. It can take almost a second to reorient attention from one item of interest to another, even under ideal, controlled laboratory situations. Generally speaking, older adults require more time to orient attention from one location to another.

Attention is captured by highly salient events in the environment, and other stimuli will not be processed during this capture of attention. Older adults tend to be more affected by salient events. Saliency is given to such stimuli as flashing, high-intensity lights and to stimuli that appear to pose an immediate "threat." Clearly, in designing for older adults, it is critical to require the smallest possible number of things to search through to perform a task. It is also critical to remove extraneous information that might "capture" attention (such as blinking display elements on a Web page). However, older adults can successfully take advantage of cues that are specifically designed to capture attention.

Spatial cognition

Some tasks require a performer to develop and reason about visual images using external cues that do not directly develop that "image in the mind." An example is translating directions and information abstracted from a two-dimensional map into an image of three-dimensional space through which one can traverse. The maintenance and manipulation of visual images involves spatial cognition. For example, in a configural learning task, people are required to combine spatial and temporal information into a representation that they can compare to a new perspective of a scene. Young adults outperform older adults on this task, especially when an unfamiliar location is tested. Age differences are also observed in tasks that require memory for object locations and the development of a sequence or route. In addition, age differences have been observed in the segmentation, integration, and transformation of spatial information. Decline in spatial ability has been shown to be predictive of proficiency in performing computer-based tasks.

Understanding written and spoken language

Linguistic representations are those based largely on verbal descriptions of situations. For example, in reading a story, individuals often develop a linguistic representation of the events in the story. An analysis of word-by-word reading times suggests that young and older adults may develop different linguistic representations during reading comprehension. Research suggests that older adults are storing smaller "chunks" and must do more frequent integration. Working memory limitations have been implicated as the source of age-related differences in peformance in various linguistic tasks, such as understanding natural language and processing and producing syntactically complex speech. Older adults also have more difficulty in comprehending language when inferences are required. That is, if connections between ideas are not made explicit, an inference must be made; such inference generation may be reliant on working memory, which is perhaps why older adults have more difficulties. If older adults can rely on their semantic memory base, language comprehension is improved. From a design perspective, familiar terms and labels should be used, and connections between concepts should be made explicit.

Procedural knowledge

Procedural knowledge is knowledge about *how* to perform activities. Procedural knowledge varies along the dimension of automaticity, from knowledge that is executed almost without thought (e.g., balancing on a bicycle) to explicit but well-practiced routines (e.g., doing long division or following a recipe). Older adults have difficulty in developing new automatic processes (conceptually, as in developing new habits) in some domains. However, in terms of tasks and activities "automatized" prior to senescence, there is much support for the proposal that these automatic behaviors remain intact. *Procedural skills* refer to overlearned procedures that are to be executed under conscious control. Algorithm application has been shown to be age-insensitive if sufficient practice is provided.

From a design perspective, it is important to make the same actions (e.g., starting a computer browser) consistent across different systems and tasks. Also, in designing training or instruction, it is critical to examine the to-be-trained activity for consistent elements of the task. These consistent elements can then become important training or learning modules. It is important to also keep in mind that older adults seem to have more difficulty as compared with younger adults when required to inhibit previously well-learned procedures. So, in designing something new, it is important to guard against the requirement to inhibit well-learned procedures or to ensure extra learning time to unlearn the old procedures and learn the new procedures. Finally, it is important to note that previously well-learned procedures may reappear in behavior when individuals are under stress or faced with multiple task demands. This is another reason to guard against designs that are inconsistent with past procedural knowledge.

Multiple-task demands

Task control involves certain aspects of tasks that interact with the basic psychological demands previously discussed. Issues of task control include speed demands and multitasking. Research results generally demonstrate a slowing of response as a function of age. Moreover, it is assumed that as the complexity of the task increases, the degree of slowing will increase as well. Although this generalization may not be strictly true, older adults are proportionately slower, from an average sense, on more complex tasks. This is true primarily for tasks that take several seconds or minutes to complete.

Virtually all complex tasks can be logically broken down into subtasks. Whether individuals break tasks down into subtasks psychologically probably depends on the extent to which the different subtasks can be performed in series. In many tasks, the different components are inextricably linked. When younger and older adults are required to perform more than one task at a time (such as driving and looking for street signs) it is generally true that older adults performed more poorly doing two tasks at once as compared to their younger counterparts. Older adults do not perform as well as young adults in dual-task conditions, and the magnitude of the age difference increases with the magnitude of task difficulty. However, in some situations, where tasks are relatively simple, older adults perform as well as young adults. When a design that is introduced requires older adults to perform novel activities, it is critical not to require the combined performance of tasks or components of tasks.

Control of movements and movement speed

A large body of literature shows that as people age, their control of movement gets worse. Generally, older adults take longer than younger adults to make similar movements, and the movement is less precise. Such difficulties occur across a wide range of activities, from difficulty in using a computer mouse to position a cursor on a computer screen to movements related to driving an automobile. This age-related difference in performance can be a major impediment to activities performed by older adults and must be considered by designers. A rule of thumb to estimate movement times (and performance of novel tasks in general) is that, on average, older adults will be approximately one and a half to two times slower than their younger counterparts.

Why are older adults slower and more error-prone when it comes to movement control? Laboratory research established the source of the age-related performance decline as a combination of (1) poorer perceptual feedback, (2) increased "noise" in the motor pathway, and (3) strategy differences in approaching the task.

This information is important and can be used from a design perspective. For example, knowing that movement control is less precise and slower, a straightforward approach to making a computer mouse interface easier to use for older adults includes simply implementing software changes in the gain and acceleration profiles that translate mouse movement into cursor movement. All current computer systems have software that allows a user to adjust the gain ratio to customize cursor-positioning performance. This is a cost-effective way to partially compensate for age-related differences in movement control in this domain.

Summary of review

This chapter was intended to provide a brief overview of perceptual, cognitive, and movement control factors that should be understood in designing for older adults. Not all capabilities decline with age, nor do some older adults show age-related declines. However, in general, some factors show age-related declines whereas others remain intact. Designers must compensate for declines and capitalize on abilities. What follows is a summary of the key findings discussed in this chapter.

Perception

- Taste and smell show age-related declines but only minimally.
- Changes in kinesthetic sensitivity make older adults more susceptible to falls and more variable in sensing movement, touch, and body position.
- Auditory declines are common, especially for older men and especially for high-frequency sounds.
- Vision declines for most older adults; visual acuity declines begin to be noticeable around age 40.
- Glare is more problematic for older, relative to younger, adults.
- Other aspects of vision also show age-related declines: dark adaptation slows, breadth of visual field is reduced, visual processing speed is slowed, and perceptual flexibility declines.

Cognition

- Working memory (i.e., the ability to keep information active) declines with age.
- Semantic memory (i.e., acquired knowledge) does not decline with age.
- Prospective memory is remembering to do something in the future. If the task is time-based (do this in 1 hour), older adults show deficits in performance. If the task is event-based (when the alarm rings, take a pill), age-related differences are shown to be minimal.

- Both selective visual attention (i.e., searching a visual display) and dynamic visual attention (reorientation of attentional focus) show age-related declines.
- Older adults can benefit from cues to orient and capture their attention.
- Spatial cognition (i.e., maintenance and manipulation of visual images) declines with age.
- Language comprehension is intact if older adults can capitalize on their semantic memory; impairments are observed when inferences are required and working memory is overloaded.
- Procedural knowledge is knowledge about how to do something. Well-learned procedures are maintained into old age and, in fact, are difficult to inhibit. Older adults are slower and less successful at acquiring new procedures, relative to younger adults.
- Older adults, on average, process information more slowly than do young adults: age-related differences increase with task complexity.
- Older adults do not perform as well as do younger adults when required to coordinate multiple tasks.

Movement control

- Older adults respond more slowly than do younger adults. In general, an older adult will take between one and a half and two times longer to respond than will a younger adult.
- Movements made by older adults tend to be less precise and more variable than those made by younger adults.

Guidance for design

In our review, several evident themes point to the sources of age differences in performance at a variety of levels. First, working memory appears often to be the limiting factor in performance (e.g., speech comprehension, manipulating quantitative representations). Pragmatically, such working memory limitations are sometimes reduced with practice (e.g., consistent practice on memory-search tasks), through the training of strategies (e.g., using external memory aids for planning purposes), or through the provision of environmental support to reduce working-memory demands. Environmental support (putting required knowledge into the world rather than forcing memory for that knowledge) has been suggested as a means of minimizing age-related differences in a number of contexts. In an attentional search task, the provision of cues directing attention to a spatial location in a display is a form of environmental support.

Our review identified processes that are important for task performance, pinpointed psychological sources of overall performance and learning decrements, and provided prescriptions for designing systems that overcome general or age-specific information-processing problems that hinder maximal task performance. Hence, we have outlined the foundation for principled task decomposition. The task decomposition identifies the psychological components necessary for novice and skilled performance and provides the principled approach to possible age-dependent remediation. In essence, it forms the foundation for the principled approach to age-specific design. In Chapter 3, we discuss the process of task analysis and using that analysis in product design.

The review of age-related effects on cognition leads to fundamental design guidelines. These design guidelines are emphasized in the chapters that follow. For example, it is important to design to limit demands on working memory and attention. One should also design to make use of previous experience. Generally, people perceive and respond rapidly to things that they expect on the basis of their experience. People generally respond much more slowly to those things that are unexpected as compared to things that are expected. One of the roles of the designer is to understand, predict, and capitalize on what people will expect. Another role of the designer is to understand that, when people are faced with using a novel product or experiencing a new environment, they will try to make their task manageable by relating what is new to what they already know. If design does not capitalize on relevant semantic memory, problems can and often do arise.

When experience cannot guide action, it is perhaps easy to see why it is critical to guard against information overload and to guard against inappropriate actions, because people thought they knew what to do based on their experience. Indeed, this discussion should point out that it truly can be difficult to interact with new products.

Recommended readings

Birren, J.E., and Schaie, K.W. *Handbook of the Psychology of Aging*. 5th edn. San Diego: Academic Press, 2001.

Craik, F.I.M., and Salthouse, T.A. *The Handbook of Aging and Cognition.* 2nd edn. Mahwah, NJ: Erlbaum, 2000.

Fisk, A.D., and Rogers, W.A. *Handbook of Human Factors and the Older Adult.* Orlando: Academic Press, 1997.

Park, D.C., and Schwartz, N. *Cognitive Aging: A Primer.* Philadelphia: Psychological Press, 2000.

Chapter 3

Guiding the design process

There is evidence that people have difficulty operating the entire array of consumer products. Although one may expect that users would attribute their difficulties to the technology, in fact users tend to blame themselves for the problems that they encounter. The root of the problem more often than not lies in the design process. The designer often unknowingly incorporates degrees of complexity into devices, interfaces, and instructions that create imbalances between the demands imposed by these products and the mental and physical resources at the disposal of the user. There are some cases in which designers may actually produce designs that underuse people's resources, and this also is an undesirable outcome. However, under the present technological and economic climate, which is resulting in a proliferation of technological innovations at very reasonable costs, it is more often the case that people, especially older product users, will be left overwhelmed and dissatisfied.

To minimize the problems people encounter in using products, more than ever it has become necessary to apply a systematic procedure to the design process. Key to such a procedure is the timely application of various methods at different stages in the design process. The procedure currently advocated is generally known as *user-centered design*. This type of design process is consistent with a human-factors approach to design. Essentially, it incorporates user requirements, user goals, and user tasks as early as possible into the design of a product, when design is still relatively flexible and when changes can be made at least cost.

What makes a product usable?

The issue of what makes a product usable can be better appreciated by initially addressing the issue of what makes a product or device useful. The usefulness of a product can be considered from two standpoints: its utility and its usability. A device's utility considers whether the functionality provides what is needed, whereas usability concerns how well users can access that functionality. From this perspective, the perceived utility of a device is fundamental to the concept of usability—a very usable product may in fact not be used if its perceived utility is insignificant.

Five components important to usability are learnability, efficiency, memorability, errors, and satisfaction. The learnability factor concerns how easy it is to learn to use the device, whereas a device that is efficient to use implies that the product should allow users to achieve their intended objectives within a reasonable amount of time. Also, efficiency will often imply the capability for matching the functions provided by the product with the needs of the users to produce acceptable product performance without inducing frustration, fatigue, and dissatisfaction.

The memorability component of usability indicates that a device should be easy to remember, thereby minimizing the effort to relearn how to use the device after periods of nonuse. Errors are generally much more easily identifiable from products with computer interfaces, where error messages are signaled to the user. However, in a broader context, errors can be construed as user actions that do not accomplish the desired goal. In any case, errors resulting from interacting with the product should be minimal and, if they do occur, the user should be able to recover from them easily. Finally, the satisfaction component addresses the pleasantness of the experience the user has in interacting with the product (i.e., the experience should be satisfying).

Although it is important to consider the perceived utility of a device and each of these usability components in designing for all potential users, these usability factors will require special attention from designers when considering older users. In particular, older users may have difficulty in learning to use a device, especially if the instructions overload the user's working memory and thus make it difficult to integrate information effectively (see Chapter 2). Also, older users may not benefit as much as younger users from the transfer of knowledge resulting from the use of other similar technologies, given general findings that older people are less apt to interact with technology. In general, memory-related problems common among many older people make it more imperative that product procedures are easy to remember after periods of nonuse. Placing the burden on older users to relearn repeatedly to use a product will likely frustrate them to the extent that they will not use the product. Furthermore, these negative experiences are likely to affect adversely the adoption of new technologies in the future, creating a downward spiraling effect that is a lose–lose situation for both designers and users, especially if the product has the potential to enhance the quality of life of older persons. Similarly, designing products that make it easy for users to make errors from which recovery is difficult will likely increase the frustration levels of older people to the extent that they become disillusioned with technological products.

Measuring usability components

Measuring usability will, in some instances, require some understanding of the methods that constitute the user-centered design process. These methods are discussed below and form the core set of tools and knowledge of which designers should be aware and with which, preferably, they are familiar. The importance of these tools at different stages of the design process and of the resources they demand, which can impact the feasibility of meeting product design constraints, are also discussed. The purpose of this section is to convince designers that components of usability can in fact be measured and to provide an overview of how one can go about obtaining such measures. It is assumed that the designer will have access to test participants and a testing environment.

Measures of learnability attempt to capture indications of the initial ease of learning. A basic measure is the time it takes users who are unfamiliar with a product to reach a specified level of proficiency in using it. The measure can be chosen to reflect the degree to which a specified task is completed successfully or the extent to which a task (or set of tasks) is completed by a specified time. To measure efficiency, it is necessary to obtain a representative sample of users who are reasonably experienced with a device. One can then measure the time it takes for them to perform various tasks that are typically performed on that device.

Measuring memorability should be confined to users who do not intend to use the device frequently. One approach is to have a test participant return to the testing environment at some future time after learning of the device and then to measure the time needed to complete a set of tasks previously learned. Alternatively, users can be asked to recall various procedures regarding device use after a test session with the device. In this case, depending on the product, the user may be able to rely on visual cues from the device to recall features or procedures.

Errors can be categorized in a number of ways, and the number of each type of error can be counted in assessing usability. Errors that are immediately detected and corrected are generally differentiated from errors that are more troublesome for users to diagnose or catastrophic in the sense that they stop the device from functioning. A mode error, another important category of error, occurs when users cannot achieve the task objective owing to the inability to recognize that the product is in a mode different from the one necessary for the product to function as intended. Other error categories often chosen are the omission of critical steps, the substitution of incorrect steps, and the execution of task steps in an incorrect sequence. Distinctions between "slips" (e.g., in which a user intends to activate one button but inadvertently presses another button) and "mistakes" (e.g., in which the action defining the error was intended but inappropriate) are also useful.

Subjective satisfaction with a device is usually measured by short questionnaires after a testing session. Users are typically asked to rate their degree of agreement with a number of statements on a scale from 1 to 5 (wherein 1 = strongly disagree, 2 = somewhat disagree, 3 = neither agree nor disagree, 4 = somewhat agree, and 5 = strongly agree). A typical statement might be, "Trying to set the clock manually on this DVD player was a very frustrating experience." The problem with using such instruments, especially when users are being paid to participate in a testing study, is that people tend to be polite in their responses. To mitigate this problem, it is recommended that some questions be included with *reverse polarity*, whereby agreement corresponds to a negative rating. Generally, these ratings are more meaningful when they are compared to ratings of different versions of the product or in comparing different populations of users, such as younger and older persons, regarding the same product. This enables ratings to be interpreted on a relativistic rather than on an absolute basis.

Methods in user-centered design

Checklists and guidelines

Checklists and guidelines constitute lists that serve to ensure that a number of issues have been considered. The problem with this method is that its degree of coverage will depend on whoever constructs the list. Likewise, the assessment of whether a product's design feature violates a checklist item or guideline may depend on the assessor. This method is usually confined to tracking very fundamental design issues, such as determining whether a product feature is too small to be detected or activated. In designing for older populations, it is critical that these lists reflect cognitive and physical limitations associated with older users (see Chapter 2). A portion of a checklist or guidelines involving push buttons for a stereo is presented in Table 3.1.

Layout analysis

Layout analysis concerns the application of various principles of display design in considering how to group or locate the functional elements of a device. Most human-factors texts (see Recommended Readings) provide a listing and discussion of these principles. Three commonly used principles are frequency of use, sequence of use, and importance of the functional element. The frequency-of-use principle states that functional elements that are frequently used should be grouped together; similarly, the sequence-of-use principle states that functional elements that are used in sequence should be

Table 3.1 A sample section of a checklist on push buttons for a stereo

- To accommodate older users with tremor and arthritic conditions:
  - For buttons with medium to high travel (1.0 mm–8.0 mm), use a low force (0.3 N–0.5 N); if a higher activation force is required, the activation force should be < 2.0 N, and the travel should be < 0.8 mm.
  - If inadvertent activation is a concern, use a medium to high activation force (1.0 N–3.0 N) with a travel not exceeding 1.0 mm.
  - Avoid combinations of low force (< 0.35 N) and low travel (< 0.2 mm).
- Ensure that there is adequate spacing between buttons to prevent inadvertent activation of controls.
- Ensure that surfaces have sufficient frictional resistance to prevent finger sliding.
- Ensure that there is proper illumination for activation under conditions of dim lighting.
- Arrange buttons to reflect sequence-of-use or group by related functions.
- If space is available, use text and pictures to represent the function of the button.
- Use buttons that provide (sufficient) tactile and audible feedback to signal that they have been activated.

grouped together. The importance principle addresses the need for making important elements easily detectable and accessible and the need for grouping certain elements together (e.g., when the inadvertent activation of one element makes it imperative that a deactivation switch be located near that element).

Although these principles are relevant to all users, they may be applied differently to different user populations. In some cases, older users may seldom use an element frequently used by younger users. Similarly, designers should be aware of the possibility that older users may attribute to functional elements of a device relative degrees of importance not consistent with the assessments of younger users. Consequently, in applying principles of design to older adults, it is important to consider the objectives, tendencies, and preferences of older users. Much of this information can be obtained from interviews, focus groups, questionnaires, and observations (as discussed further).

Observations

An obvious way of becoming informed about problems that users may encounter is to observe the users interacting with a device and to record the observations. The observer should try to be as unobtrusive as possible to foster realistic user interactions with the product. However, there may be times when interrupting a user to clarify an action is warranted. If the possibility exists for the intrusiveness of the observations to distort the user's interaction with the product, more sophisticated methods of monitoring the user's activities may be required (e.g., videotape, one-way mirror, automatic capture of keystroke data). Observation bias is also a concern; thus, designers should attempt to focus on user actions that are not necessarily consistent with what they expect such user actions to be.

Even more than with checklists, there is tremendous variability in how one chooses to organize an observational study, especially with regard to what data to collect. Typically, one provides instructions to a user concerning the task steps that the user is to perform, and the observer records steps or sequences of steps that are problematic for the user and notes any concerns that should be clarified during a debriefing session. For example, if a user is asked to set a new time on a microwave oven clock, the observer may note that the user demonstrates an ability to select the correct numbers but an inability to translate these inputs into a new clock setting.

With older users, it is not advisable to disrupt their momentum to ask questions, as this may make it harder for them to reorient themselves to the task.

Interviews

Different approaches to interviews can be taken, ranging from highly unstructured methods that elicit free-form discussions to highly structured methods whereby a predetermined sequence of questions is posed. In interviewing product users, an approach somewhere between these extremes is recommended. The interview should be sufficiently focused to capture important situational contexts but also flexible enough to allow for tangential exploration of design issues and more in-depth comments that can serve as useful anecdotal evidence for designers. Checklists and guidelines can often serve as the basis for such interviews. In an automobile, checklist and guideline items that address the organization, visual clarity, and functionality of the controls and displays of an advanced audio system can be used as a basis for a semistructured interview. This interview may determine that older adults in particular had problems finding the button for ejecting CDs from the six-CD changer in a case in which the eject CD button was located diagonally across the audio display from the load CD button. Although the designer was concerned about placing these two buttons too close together, owing to the possibility for confusing ejection with loading, the diagonal solution was probably too extreme. In addition, the interview may determine that older users might benefit from less clutter on the digital audio display and from a larger rotating volume-control button, which would afford easier control during driving. The interview provides the opportunity not only for exploring issues in particular contexts, such as the inadequacy of a control during driving in traffic or on the highway, but for exposing problems in other areas, such as instruction manuals.

Interviews should be conducted as soon as possible after user interaction with the product to minimize possible distortions or forgetting of opinions. In conducting the interview, it is important that the interviewer remain neutral by avoiding any tendency to agree or disagree with the user. Questions designed to evoke yes or no responses should be avoided, as these responses have limited diagnostic value. Ideally, interviews should capture best- and worst-case experiences with the product.

Questionnaires

Questionnaires allow users' feelings about a device to be quantified using scales, such as a five-point scale that ranges from "strongly agree" to "strongly disagree." Although a wide variety of different rating scales for responses can be chosen, it is important that the rating scale be consistent throughout the questionnaire. The items on questionnaires can be grouped to handle different aspects of the product assessment. They can also address general feelings about the device through such questions as "I felt very confident when I used this product" or "I felt that the displayed information was too cluttered." As with interviews, questionnaires should capture best- and worst-case experiences with a product, and should be conducted as soon as possible after user interaction with a product. However, unlike interviews, questionnaires cannot be modified on an ongoing basis. It is, therefore, essential that questionnaires be subjected to pilot testing and considerable scrutiny before they are administered. Careful consideration must be given to questionnaire language to ensure correct interpretation; to its content to ensure that it is capturing the intended issues; to its format to ensure that it is not frustrating or confusing to negotiate; and to its length to ensure that it is not too taxing. These considerations are especially important for older users. In general, older people are often familiar with questionnaires and may feel more at ease assessing a product through a questionnaire than verbally. The advantage to the designer of having responses quantified, which is possible with questionnaires, is that it can provide a basis for drawing conclusions. Analysis of questionnaires may allow the inference, for instance, that older people have more problems than do younger people in finding information or that most users are confused about how to shut off a device.

Focus groups

Focus groups are essentially discussion groups that are composed of approximately 6 to 12 users or potential users of a device and are brought together to discuss user needs, feelings, experiences, and opinions and to generate ideas and recommendations. The group can be brought together during preliminary stages of design, to discuss past problems or issues in using the type of product that is being conceptualized or is in development, or can be assembled after use of a prototype design. A moderator or group leader is responsible for maintaining the focus of the group and usually relies on a script containing issues for discussion and guidelines for the kinds of information to be collected. The advantage of focus groups is realized when the group interactions foster ideas that would not likely have been offered in one-to-one interviews. Thus, it is necessary for the moderator to avoid having one or a few participants inhibit others; having focus groups composed of a homogeneous sample of users usually minimizes this problem. If younger and older adults are to be interviewed, it is best to conduct separate group interviews, with younger adults in one group and older adults in another group. Homogeneity within groups tends to lead to more spontaneous discussion. Data analysis can range from a report that reflects the overall attitude of the group or that highlights the various issues and concerns to suggestions for improvement.

Using a sample of older users is recommended as a means for *exposing hidden problems* with products, especially in their use in particular contexts, and for generating a variety of suggestions from this group of users. If the design intent involves a product or system to be used at work by older workers, the focus group should be limited to older people still engaged in or actively pursuing work activities.

Verbal protocols: thinking aloud

In using the verbal protocol method, a test participant continuously "thinks aloud" while interacting with the product, with the emphasis being on what the user is doing and why it is being done. This method circumvents reliance on rationalizations that are offered afterward concerning problems the user encountered, and it provides excellent qualitative data that can be integrated into reports by designers for supporting arguments for redesign or product embellishment.

Designers, however, must be cognizant of two important issues when using this technique. First, not all people are comfortable verbalizing their thoughts while occupied in some activity. Fortunately, test participants typically consist of nonexpert users who will likely have much more to say about their interaction with the product than would expert users (whose degree of expertise makes it difficult for them to articulate their interactions with the device). However, there are people who are not good verbalizers; that limitation requires the designer-experimenter to use such prompts as "Please keep talking" to ensure that the user continues talking. The experimenter, though, must be careful to avoid questions that interfere with the user's interaction with the product. Providing a warm-up thinking-aloud exercise is recommended to minimize the possible discomfort associated with thinking aloud as well as the tendency not to verbalize. As an illustration of this approach, the user can be asked to think aloud while searching for the number of an airline by using the telephone book.

A second major concern with thinking aloud is that the process of verbalizing thoughts may redirect users' attention and problem-solving behavior in a way that changes their interaction with the system, often for the better. To minimize this possibility, the balance of the thinking-aloud method should be directed toward what the user is doing rather than toward detailed rationalizations of activities. Conversely, for older adults, that process of verbalizing their activities creates a dual-task situation, which may cause them to perform less skillfully than they would if they had not been asked to think aloud. This issue may be addressed by allowing the person to perform the task first, without thinking aloud, and then to perform the task a second time while thinking aloud.

Task analysis

How does a designer determine how the functional elements of a device are used to meet a user's goals? A method that is frequently used by human-factors specialists is task analysis. This method decomposes tasks the user performs with the product or device into steps that ultimately provide information concerning the requirements for accomplishing the task objectives.

A particular type of task analysis method that has been found to be useful in product design is hierarchical task analysis (HTA). In using HTA, the task the user must perform is decomposed hierarchically into goals, the plans for meeting these goals, and the operations for carrying out these plans. In using a videocassette recorder (VCR), a goal may be to program the taping of a future television program. The plan for meeting this goal may consist of (1) turning on the television; (2) setting the appropriate channel; (3) turning on the power to the VCR; (4) obtaining the menu display for taping future programs; (5) selecting the correct option from this menu; (6) entering the appropriate date, time, and station information; and (7) executing and confirming the action. In specifying a plan in an HTA, the order in which the plan is carried out is usually specified, as it is possible to meet the higher-order goals in different ways. Finally, for each step of the plan, the operations needed would be specified in the order in which they need to be performed.

Although graphical flow-chart formats are often used to depict an HTA, tabular formats are recommended to enable the designer to include additional columns that can contain useful information related to operational steps. Examples of such information include:

- The type of action or behavior required by the user
- The potential for errors associated with these actions (e.g., the possibility for confusing AM with PM or inadvertently executing an action)
- Opportunities that exist for recovery from these errors
- Excessive demands imposed by that operation (e.g., memory, visual, or motor demands)
- The potential for injury or the creation of hazardous conditions.

A task analysis performed early in the design process is often termed a *preliminary task analysis* and should be differentiated from a task analysis performed when a prototype of a device exists. Task analysis is extremely important for early input into the design process, especially in designing for older adults. Its ability to identify information needs; visual, auditory, and tactile detection requirements; demands for focused attention and for retaining information in memory; the time necessary to react to signals; and such physical requirements as digit manipulation and required forces provides a starting point for pinpointing problems that older users potentially face.

At the later stages of product development, users interacting with a product or system can be observed for various tasks or goals that they would like to accomplish. Task analysis at this stage can be used to validate the preliminary task analysis (i.e., it can be used to determine whether certain task demands are still a factor). However, now users can be asked questions concerning why they did a certain action or how they went about accomplishing a particular step of a plan. Greater insight into dependencies between task steps and, in general, into difficulties in accomplishing objectives can now be obtained.

Finally, it should be noted that task analysis is a powerful tool that can be applied not only to a user's interaction with the device itself but to instructional manuals or any other "tool" the user requires for performing the task. It is not uncommon in industry for task analysis to be used to analyze and redesign written operating procedures that workers follow to perform a task. When applied to older users, task analysis may determine that the instructional manual contains textual information that is difficult to read and comprehend and lacks diagrams or pictures that would allow the user to identify important functional elements associated with the device. Similarly, task analysis applied to a multimedia training package may expose excessive demands placed on older users originating from their need to process simultaneously information from multiple sensory modalities.

Safety and environmental analysis

Some products are potentially hazardous in ways not foreseeable by users. For such products, older users may not detect or interpret warnings as readily as younger users might, and may have physical limitations that could increase the likelihood of creating hazardous conditions. Task analysis—in particular one that addresses such cognitive demands as the need for discriminating between warning indicators, interpreting messages, or requirements for focused or divided attention—is essential for predicting the possibility of over-loading or confusing users. The HTA approach described earlier is particularly well suited for this purpose. For each step of the analysis, the possibility for different types of errors can be assessed (distinctions between different types of human errors can be found in human-factors texts or in works addressing human reliability; see Recommended Readings). Using a tabular HTA format, one can also include a column that addresses the consequences of an error and the possibility for a user to recover from such an error. An environmental analysis can supplement the safety analysis by providing a more detailed analysis of the contexts within which a user interacts with the device or system. This analysis can help the designer to determine whether a hazardous condition or human error may or may not result in an adverse outcome for the user. For example, if the product were an automobile's voice-operated navigational system, it would be misleading to evaluate an older driver's use of this device while sitting in an unmoving vehicle.

Usability testing issues

There are a host of usability testing issues that may have to be considered, depending on the rigor the product designer is willing to accord to the usability testing process. For the most part, these issues apply equally to older and younger users, although older users may in some instances demand a more careful consideration of these issues. Many of these issues can be considered within the context of a "usability test plan."

Statement of goals

The initial item that a usability test plan should include is a clear statement of the goals of the test. Examples of such goals are determining whether the user can find certain functions without the use of an instruction manual or determining whether the user can recover from certain types of errors. Note that the determination of how older users fare as compared to younger users is a goal that should be addressed in the test plan.

Sampling and statistical analysis issues

The second item a usability test plan must address concerns sampling and statistical analysis issues. Typically, test users are selected on the basis of being representative of the people who are known to use the product in question. However, it is recommended that designers take a broader view and consider not only anticipated users of the product (as based on past marketing and sales data) but users (such as older adults) who can potentially benefit from the product. The sampling issue also must consider the level of skill the user has in using that type of product. Although it is important to test users who are unfamiliar with a product, it is also essential that users who have experience with similar products be tested. This allows a "boundary of expectations" to be established: if people very familiar with the basic product are having difficulty, it is unreasonable to expect novice older users to have much success.

Another sampling issue concerns the number of test users that will be needed. To some extent, this depends on whether statistical analysis will be performed, such as computing "confidence intervals" that can provide estimates of the average and of the variability of various measures of performance. Suppose confidence intervals are computed for the average time it takes to set a clock on a microwave oven. If the confidence interval for the group of older users includes unacceptably long average times, there is a reasonable concern that older users may become frustrated with this task and ignore it. In cases in which a "target value" for a parameter has been established, a more formal statistical procedure (termed *hypothesis testing*) can be used. In the microwave oven example, this procedure would enable an analyst to determine whether there is strong evidence that older users on average exceed the targeted (in this case, upper-boundary) time for performing this task.

When the interest is in comparing two or three versions of a product, a relatively larger number of test users may be required. In these cases, methods of "experimental design" can be applied to answer such questions as whether one of the designs is superior or whether older users perform best on a design that differs from the one with which younger users do best. These statistical methods can also be applied to subjective data from questionnaires to gauge differences in preferences. It may be of interest to determine, for instance, not only whether differences in product preferences exist between younger and older users but whether there are inconsistencies between user preferences and user performance with the product. In such cases, designers should consider why the design that is preferred by the user is not proving to be the most effective one to use.

If the intent is to have each user try different versions of a product, the designer must be cognizant of the effects of the order in which the test participant experiences the alternative designs. In the case of two alternative products, the use of "counterbalancing"—in which half the users experience the sequence reversed from that of the other half—makes it possible to control for the effects of sequence of product use. However, this approach would not handle situations in which the effect of being exposed to Design B after exposure to Design A is different from the effect of being initially exposed to Design A followed by exposure to Design B, illustrating the level of detail designers need to take into account in planning usability studies. In general, if the intention is to perform statistical analysis, the designer needs to define clearly the measures that will be collected (e.g., time to complete the task, number of times the user returned to a previous step) as well as the methods that will be used to analyze these data.

Other usability testing issues

Prior to formally testing users, "pilot tests" should be performed on a small group of users. At least one pilot participant from each targeted age group should be tested. The purpose of pilot testing is to identify and ultimately remove any problems that would undermine formal user testing. These problems might include incomprehensible instructions or questionnaires, unreasonable amounts of time allocated for testing, inability to collect certain measures, or the discovery that certain measures are incorrectly defined.

Another important consideration in usability testing is the extent and form of training that will be needed on the tasks. Some products, such as telephone voice menu systems, should probably be tested with minimal instructions, if any, to be consistent with the fact that users typically encounter these systems in the real world without any instructions on their use. Training can be imparted through manuals or face-to-face communication. For some types of testing, the interest may be in determining whether the user can learn to solve new functions, given a basic understanding or overview of the product. In other cases, the interest may be in determining how a user acquires this basic knowledge. These objectives should dictate the strategy used for imparting knowledge to users about the product.

Finally, the nature of the interaction between the experimenter and the participant during training should be resolved prior to testing. Many participants, especially older users, are likely to ask questions as they confront various types of difficulty or uncertainty during their interaction with a product. A protocol should be established governing the rules associated with the kind of help an experimenter could offer. This provides a degree of standardization across test participants that helps to ensure that they all receive the same kind of information.

Selection of methods

As implied earlier, various methods are available to designers, and ultimately it is the judgment of a designer that decides which method or combination of methods should be used and how refined these methods or approaches should be. If a number of stages in the product design life cycle can be reasonably identified, it is possible to estimate the stages at which each of the methods is applicable. Table 3.2, Panel A, presents six general design stages. In Panel B we provide estimates concerning how resource-intensive the different methods are, where resources reflect costs associated with development, administration, and data analysis associated with the method. For some methods (such as task analysis), use rather than administration is a more accurate description of the activity. It should be kept in mind that these estimates could change dramatically, depending on

Table 3.2 Stages of design

A. *Design stages*

Design stage	Description
1. Conceptual design	The idea for the device is considered, and many implementations of the design are still viable.
2. Formalization	The idea becomes more formalized, and there is a corresponding reduction in the number of feasible design solutions.
3. Design	A design solution is derived, and the plan for developing the product is devised.
4. Prototyping	A prototype of the product is developed for analysis.
5. Commissioning	The final design solution is implemented, and the product enters the marketplace.
6. Operation and maintenance	The focus shifts to supporting use of the product in the marketplace.

Source: Adapted from Stanton and Young (1998).

B. *Estimates of resource-intensiveness of methods*

User-centered design method	Design stages	Development	Administration	Data analysis
Interviews	1–6	Low to moderate	Moderate to high	Moderate to high
Questionnaires	1–6	Moderate to high	Low to moderate	Moderate
Focus groups	1–6	Moderate	Moderate	Moderate to high
Checklists and guidelines	3–6	Moderate	Low	Low
Task analysis	3–6	High	Moderate	—
Safety and environment analysis	3–6	Moderate to high	Moderate	—
Layout analysis	4	Low to moderate	Moderate	—
Observations	4–6	Low	Moderate	Moderate
Verbal protocols	4–6	Low	Low to moderate	High

Note:
Low, moderate, and high refer to the resource intensiveness for each method as a function of costs associated with development, administration, or data analysis.

a number of considerations, including the type of product and the depth to which a designer-analyst would like to apply the method. Investment of resources is categorized as low, moderate, or high for each method as a function of development, administration, and data analysis efforts.

Conclusion

In this chapter, a number of methods and issues involving user-centered design were discussed. In principle, user-centered design pertains to design for all users. In reality, the most useful method for determining design for older users will depend on the product. Familiarity with these techniques can provide designers with the knowledge necessary for deciding which methods to select when older users are a concern. It can also provide the insight necessary for determining how to tailor these tools toward identifying problems older users may face in interacting with devices and for determining the potential solutions to those problems. Informal examples were pointed out throughout the chapter to help to orient designers toward achieving these goals.

It is critical to remember that the designer, the engineer, and so on are *not* the sole arbiters in determining effective usability of a product. Representative users performing representative tasks within representative contexts must be the arbiters of usability.

Recommended readings

Kirwan, B., and Ainsworth, L.K. *A Guide to Task Analysis*. London: Taylor and Francis, 1992.

Nielson, J. *Usability Engineering*. Cambridge, MA: Academic Press, 1993.

Rubin, J. *Handbook of Usability Testing: How to Plan, Design, and Conduct Effective Tests*. New York: Wiley, 1994.

Stanton, N. *Human Factors in Consumer Products*. London: Taylor and Francis, 1998.

Whitley, B.E. *Principles of Research in Behavioral Science*. 2nd edn. Boston: McGraw Hill, 2002.

Wickens, C.D., Gordon, S.E., and Liu, Y. *An Introduction to Human Factors Engineering*. New York: Longman, 1998.

Part II
Design guidelines

Chapter 4

Improving perception of information

This chapter is divided into sections on vision and hearing. Our goal is to provide a brief summary of some of the sensory and perceptual changes that occur with age and to examine the implications of those changes for the design of products and environments. Our aim is to develop design principles and recommendations that can enhance the likelihood that older adults will be able to interact successfully with technical systems.

For example, many older men have difficulty in hearing sounds in the 8,000+ Hz frequency range, even at very high sound levels (90 dB). Hence, it would be foolhardy to signal a dangerous situation using that frequency range. Another example is the decline in sensitivity to illumination with age. Under low light conditions, an older eye admits approximately one-third of the light to the retina admitted by a younger eye. Hence, there is particular need to ensure that light levels in the homes of older adults are adequate to illuminate workspaces (e.g., bathroom counters where they might read medication labels or kitchen areas where they might read food or cleaning solution labels).

Our focus is on areas of particular relevance for design for older adults and only on visual and auditory perception as it is most relevant to the other topics. Again, it is worth stressing that interventions that help older adults may also be useful for younger adults.

Issues in aging and visual perception

There is considerable diversity in visual capabilities within the older adult population, although in general, prevalence of visual impairment accelerates after age 65. The diversity is due in part to variability in aging processes and also to the increased use of assistive devices by some older adults to compensate for age-related changes. Finally, there is an increased frequency for surgical interventions that modify the visual system of older adults, such as cataract operations to eliminate opacities that develop in the lens and elective surgery to reshape the cornea. One of the most prominent age-related changes in vision is the decreased transmission of light to the receptors in the back of the eye. This occurs in part because of decreased ability to dilate the pupil as widely and to yellowing of the lens (also reducing color discrimination for short wavelength hues) and increased scattering of light in the optical media between the cornea and the visual receptors at the back of the eye. Cataract surgery can improve light transmission. However, there are also changes in the visual cortex that limit visual acuity. Hence, it is not always possible to correct visual acuity to 20/20 vision. Another prominent change with age is loss of focusing power in the lens, making it difficult to change its shape to view near objects. There are also declines in the ability to judge depth and to judge motion (particularly, time to impact in gap detection situations).

A useful way to conceptualize these age-related changes, initially advocated by Alan Welford, is to view an older adult's perceptual system as an information channel that is noisier than that of a younger adult. Thus, for any strength of external signal, the perceptual system's output—the signal-to-noise ratio—is likely to be lower in older than in younger adults. Designers need to consider classes of interventions that boost the signal strength of messages and that reduce the sources of noise for the system. Ways to boost "signal strength" include increasing the size of visual objects (e.g., font size, icon size), their brightness, and their contrast. Ways to decrease "noise" include isolating messages from other message channels (e.g., avoid putting objects in the periphery that could attract attention, such as advertisements on Web pages), and maintaining consistent positioning of target items (e.g., location of help information). See Chapter 7 for specific advice for interface design.

This advice is not always easy to follow. For instance, the simple advice to increase luminance levels can lead to other problems. Sudden changes in illumination from point to point may overwhelm the adaptation capabilities of older adults. They require approximately three times longer than younger adults to reach maximal sensitivity when moving from bright to dim light conditions (e.g., leaving a brightly lit room and going outdoors at night). Simply increasing light intensity may not be an effective solution to improving legibility of print because there is also increased scattering of light through the older eye. This scattering increases the risk of impairment from glare sources, such as that produced by reflective surfaces (e.g., glossy magazine pages under bright light). Because vision is a long-range sense (providing information earlier than hearing or touch) and because vision plays an increased role in balance with age, improving conditions for visual perception can be of great help to the productivity, comfort, and safety of older adults. The relevance of these issues to the workplace is discussed in Chapter 8.

In our highly literate society, much of the information that we consume is in the form of printed text. Increasingly, such text is conveyed via computer-controlled monitors. In this section, we stress interventions that should improve the processing of text materials, whether displayed with light reflective sources (ink on paper) or light transmissive ones such as cathode ray terminals (CRTs) or backlit liquid crystal displays (LCDs). Such displays are increasingly deployed outside work environments and in the home (for example, in health care technologies, as discussed in Chapter 9).

Finally, designers should consider engaging alternative sensory systems by providing redundant channels for those who have severe visual impairments. An example for warnings would be using sound and vibration in addition to visual signals. Our research has found, for instance, that providing both auditory and visual text information via screen phones helped older adults to interact successfully with telephone menu systems. However, it is also important to avoid too much information that might overload working memory.

Text characteristics

Printed text (e.g., in instruction manuals for software or airport screen displays) can be distinguished by the type of font (e.g., serif or sans-serif); the thickness of the font (weight, such as normal versus bold text); the size of characters within the font set (usually measured in terms of x-height, which is the height of the character x in the chosen font); the brightness and color of characters and their background; and the spacing of text (interline spacing, number of columns; justification to the left, center, or right). Perhaps most critical is the visibility of the text, determined in part by the contrast ratio between text and background, which depends on ambient luminance for reflective displays and intensity of foreground–background elements for transmissive displays. (Specific recommendations are provided below.)

Icons versus text

Icons and other symbolic displays can be effective ways to convey information if older adults are already familiar with the meaning of an icon or symbol. For example, some icons used in current software packages are ambiguous, and the symbols give little indication of their meaning; this defeats the utility of the icon. Icons and symbols must also be easily discriminable. Research on perception of traffic signs indicates that as long as the symbols are well designed (do not require the ability to process high-frequency spatial information, that is, acuity for fine detail), they can be processed as well as or more effectively than text messages. Icon sets (e.g., the internationally adopted set for traffic signs) become useful only after an opportunity to learn their meaning. Older adults can be expected to take longer to learn arbitrary symbol sets and to be less likely to remember them. For warnings (e.g., medication labels), such sets may not be as effective as text for older native language speakers.

Additional factors to consider

Worth keeping in mind is that not all adults have the same native language. Therefore, choice of vocabulary and content of information are important. Technical language used in many instruction manuals and help systems may be particularly difficult for older people. Also, educational attainment levels for older adults are lower than for young adults, so it is important to use vocabulary that will be familiar to all possible users.

One concern with computer-driven displays is that screens can display many different elements in a variety of attention-catching formats. For instance, flashing and scrolling text and images in the periphery are particularly problematic for older adults trying to read text because they are less able to ignore distractions. Such attention-grabbing techniques should be minimized, particularly for screens displaying critical information (e.g., warnings). Research has also shown that older adults have less effective "useful fields of view" in such situations as driving. Older adults are less likely to process events in peripheral vision as successfully as do young adults (e.g., to detect oncoming traffic in the midst of cross-traffic-lane turns).

Guidelines for visual presentation of information

General lighting guidelines

- Increase the level of illumination to greater than 100 cd/m² light reflected from reading surfaces (such as white paper). Photometers (luminance meters) can be used to assess light levels.
- Reduce direct and reflected glare by positioning light sources as far away as practical from the operator's line of sight, using several small low-intensity light sources rather than one large high-intensity light source, by shielding light sources or using diffusers on them, and by reorienting the work surface or furniture.
- Provide adjustable light sources (such as desk lamps) and use nonreflective materials on walls, floors, and ceilings. Matte surfaces are preferable to glossy ones. Ensure appropriate coverings on windows (blinds, shades) to shield work surfaces from direct sunlight.

Text

(Examples of good and bad text presentation styles are presented in Figure 4.1.)

- *Font size*: Select 12-point x-height fonts in designing for older users. A 12-point x-height means that the height of the *x* character is approximately 4.2 mm.
- *Font scaling for Web pages*: Avoid style sheets that prevent people from increasing font size with their browser software. (Some newer versions of browsers allow overriding of style sheets, but many users may be unaware of how to accomplish this.)
- *Font type*: Avoid decorative and cursive fonts (e.g., gothic); prefer either serif or sans serif fonts, such as Times Roman, Arial/Helvetica. Reading is slowed by entirely upper-case text as compared to normal text. However, UPPER-CASE TEXT attracts more attention than does lower case in mixed case situations.
- *Contrast ratio*: Try to achieve at least 50:1 contrast (e.g., black text to white background, measured from solid black and solid white areas); for transmissive displays, prefer LCDs rather than

Good: This text is a good size for seniors (12pt Times)
Bad: This text is a bad size for seniors (8pt Times)
Good: Arial is a good font to use for seniors
Bad: *A script font is a poor choice for seniors*
Good: Black text on a white background is easy to read,
as is white on a black background
Bad: Black text on a gray background is difficult to read

Figure 4.1   Examples of good and bad text displays

CRTs when screen size is held constant because of the generally higher contrast ratio on backlit LCDs. (Luminance meter readings taken near the screen on a white and black patch on a typical LCD monitor are $140/.8 = 175$, and from a CRT monitor are $71/1.5 = 47$.)

- *Color selection*: Ensure that color discriminations can be made easily. For example, avoid signaling important information using short wavelength (blue-violet-green) contrasts; text should be black on white or white on black to maximize contrast; avoid colored and watermarked backgrounds for text display areas (such as black text on blue backgrounds). Consider providing white on black text when using CRT displays for those users with significant visual impairments.
- *Motion*: Scrolling text is difficult to process and should be avoided.

Issues in aging and auditory perception

In this section, we discuss principles and guidelines to help to ensure that older adults receive needed auditory information. Our focus is on making speech more intelligible and on improving the efficacy of warning signals. We begin with a brief review of age-related audition changes that have relevance to design.

Pure-tone thresholds increase with age, particularly for high-frequency sounds beyond 8,000 Hz. Ability to hear speech (tested by isolated monosyllabic words) declines in the decade of the 50s, particularly for men. As well, masking of signals by noise increases with age, so that even if the intensity of sounds is boosted at the ear with an amplification system (e.g., a hearing aid), information may remain unintelligible because background noise may be boosted at the same time.

To compensate for losses in hearing acuity, older adults may need to use context to interpret speech. Studies have shown that they depend more than do young adults on context, such as degree of predictability of a target word. Having good structure (e.g., grammar) in spoken (and written) texts can help older adults differentially. For instance, pausing after important grammatical boundaries (phrases, ends of sentences) when speaking may be particularly helpful. Another important issue is the slower rate of processing for older adults. This has implications for the use of compressed and speeded speech.

For example, in one study of telephone menu systems, we found that older adults had more difficulty in processing menu information when the speech was compressed at 20 percent. Designers of these systems often compress speech at higher rates to maximize efficiency; however, this may place older adults at a disadvantage (e.g., they may need to repeat the menu more often). The same is true for speech rate of messages on telephone answering machines. This is an increasingly important issue as we rely on use of these systems to convey important information, such as reminders of doctor's appointments or medication reminders (as discussed in Chapter 9). Television and radio announcer speech rates for newscasts are a good standard to emulate.

The idea of a noisy communication channel is helpful in trying to understand age-related changes in auditory perception. For any signal strength in the environment, an older adult's perceptual system transmits a lower signal-to-noise ratio than does a younger adult's system. Hence, interventions should aim at increasing signal strength and decreasing the intensity of noise. Another strategy to consider is to make use of other sensory modalities to compensate for negative age-related changes in hearing.

Thresholds for sounds

Pure-tone thresholds are measured by assessing the minimal intensity necessary to detect a tone of a given frequency. Such tests are administered under ideal conditions (quiet room with headphone presentation of the signal) and with highly controlled sounds (narrow frequency bands). For older adults, there is a marked increase in the intensity needed to hear a sound with increased frequency of the sound. Most natural sounds in the world are complex, with multiple frequencies represented, although there are usually distinct fundamental frequencies or harmonics containing the highest energy. For instance, for human speech, most of the energy for vowel speech sounds is concentrated between 100 and 4,000 Hz. Nonetheless, such English consonants as the sound "ess" (*s*) exhibit much of their energy beyond this range (up to 8,000 Hz), so speech perception can be impaired in older adults for sounds in the upper frequency range. Someone with age-related hearing loss could misidentify words by missing consonants with energy mostly concentrated in high-frequency bands. Because women and children tend to have higher pitched voices, they are potentially somewhat more difficult to hear by those with high-frequency hearing impairments. Thus, male voices are preferable for public announcements.

Speech perception

Typical environments have ambient sound sources that can mask "signal" sources, such as speech or warning sounds. The range of masking for a given frequency increases for older listeners. A good solution to this problem, sometimes seen in museum settings, is to provide individual headphone sets for users; they insulate the listener from ambient noise and provide a high-quality speech stream whose volume is adjustable.

Another concern is the format of speech. Although English is a nearly universal language today, there are many varieties of accent within the English language. In the United States, northerners often have great difficulty in understanding native southern speakers, and many Americans have problems in interpreting some British accents. Most people, however, are exposed through the media to "standard English accents," sometimes termed *Midwestern broadcast English* in the United States (or *BBC broadcast accent* in the United Kingdom). For prerecorded speech, using speakers with such accents may be particularly helpful to older listeners who may have problems coping with high ambient noise. For similar reasons, computer-generated synthesized speech may be difficult to comprehend by older listeners. Good practice would be to emulate the Atlanta, GA airport subway system's use of both speech and simultaneous visual (screen) display of stop information for the train.

In built environments, sounds reach the ear from multiple routes at different time delays because of reflection from floors, walls, and ceilings. When the time delay is slight, the auditory system is able to suppress the time of arrival differences. As it lengthens, the listener hears echoes that can interfere with both comprehension and localization processes. For example, speech messages regarding gate changes and flight delays in airport environments are often subject to masking by ambient noise and sometimes to distortion by echo. Such phenomena imply that it is important in designing and testing auditory signals to consider the environments in which the signals will be heard.

Localization

Sounds arrive at our two ears with time and intensity (and phase) differences related to the displacement of the sound source from the midline position of the head (between the eyes). We localize the source (determine direction) from this information. If the sound is continuous, we can locate it by moving our heads and sampling its intensity change. However, with high-frequency and short-duration sounds, localization becomes difficult. Miniaturized devices often rely on generators or oscillators that by virtue of their small size emit most of their energy in the high-frequency part of the sound spectrum. Because many of these devices are battery-operated, the usual volume and duration of the warning signal is usually not optimal for localization. For instance, many electronic watch alarms use high-frequency short-duration beeps that make the sound source very difficult to localize. Listeners are hard-pressed to decide whether it was their watch or someone else's that sounded. In contrast, the continuous lower-frequency intermittent sounds that many commercial vehicles use to signal that they are in reverse gear (backing up) are easier signals to localize.

Guidelines for auditory presentation of information

Sound

- Permit users to adjust sound volumes. It is important to provide instructions regarding how to make volume adjustments.
- Avoid frequencies beyond 4,000 Hz.
- For warning signals, try to keep most of the energy spectrum for the signal within frequency ranges of 500 to 2,000 Hz and intensities at least 60 dB at the ear of the listener.

- Consider providing redundant information. For instance, augment warning signals by using another sensory channel, such as vibration or light. Consider providing parallel visual and auditory presentation of language (e.g., using speech recognition or closed-caption text for public addresses).
- Minimize background noise and reverberation. For example, use sound-absorbing materials on walls, floors, and ceilings. Provide wireless headphone sets to older listeners in public settings. Avoid background music during spoken language (e.g., in movie or television segments).

Speech

- For presentation of speech information, ensure adequate pauses in speech at grammatical boundaries (e.g., pause after phrases and at the ends of sentences).
- Maintain speech rates to 140 words per minute or less.
- Match voice characteristics to the situation. Prefer male voices to female voices for announcements. Prefer female to male to get attention. Avoid artificial (synthesized) speech messages that do not closely imitate natural speech.
- If sound location must be signaled with high-frequency sound sources (fundamental frequency > 2,000 Hz), use longer duration (> 0.5 s) sounds.

Using these guidelines for optimizing perception of information

Perceptual processes provide users with their initial representation of a device and enable them to monitor its changes over time. As always, *"honor thy user,"* and particularly older users, by ensuring that the demands made on their somewhat noisy perceptual systems are minimized. A good heuristic for optimizing perception of information is to increase signal strength and reduce noise sources. Another is to provide redundant channels. For instance, in the case of text, boosting signal strength involves choosing a legible font (type, size) and increasing the contrast between text and background. The latter can often be accomplished by boosting light levels. Diminishing noise involves isolating important text from its surroundings, usually by enhancing it (e.g., **putting text in bold**). In speech, increased signal strength can be promoted by regulating speech characteristics to match listener needs and by allowing users to control volume. Reducing noise usually involves controlling sources of noise (e.g., dampening sound emissions from heating and cooling systems in built environments) and choosing building materials that absorb sound effectively. When feasible, try to provide multiple channels for important information, such as speech and visual signs.

Recommended readings

Charness, N., and Bosman, E.A. Human factors and age. In F.I.M. Craik and T.A. Salthouse (eds), *The Handbook of Aging and Cognition.* Hillsdale, NJ: Erlbaum, 1992:495–551.

Charness, N., and Dijkstra, K. Age, luminance, and print legibility in homes, offices, and public places. *Human Factors* 41, 2 (1999), 173–93.

Fozard, J., and Gordon-Salant, S. Changes in vision and hearing with aging. In J.E. Birren and K.W. Schaie (eds), *Handbook of the Psychology of Aging.* 5th edn. San Diego: Academic Press, 2001:241–66.

Kline, D.W., and Fuchs, P. The visibility of symbolic highway signs can be increased among drivers of all ages. *Human Factors* 35 (1993), 25–34.

Legge, G.E., Rubin, G.S., Pelli, D.G., and Schleske, M.M. Psychophysics of reading: II. Low vision. *Vision Research* 25 (1985), 253–66.

Schneider, B., and Pichora-Fuller, M.K. Implications of sensory deficits for cognitive aging. In F.I.M. Craik and T. Salthouse (eds), *The Handbook of Aging and Cognition.* 2nd edn. Mahwah, NJ: Erlbaum, 2000: 155–219.

Tinker, M.A. *Legibility of Print.* Ames, IA: Iowa State University Press, 1963.

Chapter 5

Developing training and instructional programs

Rapid developments of technology and the diffusion of technology into most settings imply that people of all ages, including older adults, are constantly confronted with the need to learn new things at work (e.g., new software applications, new job procedures); at home (e.g., medical devices, media products); and in service environments (e.g., automatic teller machines (ATMs), self-service ticket kiosks at airports). In addition, older adults generally report that they would be more receptive to using new technologies, such as ATMs, if they were provided with training and instruction. Thus, the topics of learning and training are critical issues in considering design for older adults.

Our goal in this chapter is to provide some basic principles and guidelines regarding "best practices" for training older adults. Before we get to the details, we present a basic overview of the learning and skill-acquisition processes to provide a scientific foundation for our recommendations. As discussed in Chapter 2, older adults experience changes in many cognitive abilities, such as working memory, perceptual speed, spatial cognition, and attention. These changes may influence the manner in which they learn new things and have implications for the design of training programs.

Training materials will not be optimal if the to-be-learned material is presented in a highly paced fashion or if instructional materials are designed such that materials must be integrated across sections. Research has shown that this can be particularly true for older adults. Other important issues in the design of training programs include timing and amount of feedback, amount of practice, practice schedule, and training media. In the following sections, we focus on issues that are especially critical for older adults. It is important to note that many of these issues are critical aspects of good design for people of all ages: good design for older adults is usually good design for people of all ages.

Stages of learning

The process of learning or of skill acquisition generally involves a learner's progression through a number of stages. Although researchers have argued about the exact number of these stages and what occurs during them, we describe a three-stage process that can be applied to a wide range of learning situations. Understanding of the fundamentals of the learning process provides a basis for knowing what instructional strategies are most effective in teaching older adults and for understanding training guidelines that represent best practices.

Stage one

The initial stage of learning generally requires a reasonable degree of attention and cognitive effort as the learner is attempting to understand fundamental concepts and basic performance requirements associated with the to-be-learned material. For example, in the case of learning to use the World Wide Web, the learner must understand new concepts (e.g., what the Web is and how it can be used); various stimuli, such as buttons, displays, menus, switches, and input devices; and a host of technical terms. For complex tasks, such as learning to navigate the Web or many work tasks, providing an overview of the task is important at this stage. This knowledge provides the learner with a context for understanding why individual subtasks must be performed and how they are related.

In a study that involved training older adults to perform a telecommuting task that required servicing customers' queries and concerns through e-mail correspondence, we provided an overview of the nature of the job prior to training on specific subtasks, such as how to open and process an e-mail, how to recognize the distinctions between the information contained in different sections of the database, and how to use the interface tools required for searching and selecting information from the database. Generally, we have found that older people benefit from being provided with an overview of the purpose of a task or procedure, especially if the domain is unfamiliar. This helps them to understand not only the overall system dynamics but why certain procedures or tasks have to be performed. Sometimes, if the different parts of a task are presented in isolation, a learner may not understand the relevance to the overall task or system. For example, in setting up a new personal digital assistant, users may not understand the need to install the application software on their computer unless they understand the benefits of linking the two systems. However, the development of the proper overview information (i.e., advance organizer) is itself a challenge that requires prototyping and user testing: it is not the case that just any information provided in advance is going to improve system understanding.

During the initial stage of learning, people usually work from instructions provided by a trainer (e.g., in a classroom or work setting) or from instructions that can be presented through various media (print, video, etc.). Because older adults process information at a rate slower than that of younger adults, the pacing requirements of the training program are critical in training older adults. Generally, self-paced learning schedules are preferred. It is important to consider the characteristics of the group in training older adults in classroom or group sessions. The questions posed and the pace that is set by some of the younger learners, which may be due to a greater familiarity with the material (especially in the case for many types of technology), may intimidate some older individuals. In general, mixing younger and older adults is not advisable if age-related differences in experience levels are suspected.

The traditional classroom type of instruction involves presentation of a large amount of knowledge concerning facts and concepts imparted to the trainees prior to their performance of a task. This format requires that a great deal of information be learned. Memorizing facts and incorporating that information into previous ways of thinking may be especially difficult for older adults, given age-related declines in working memory. Thus, such approaches as part-task training (discussed later) should be considered for these types of situations.

Although a "learning-while-applying" approach that allows people to process information while they are performing the task has become a popular method for dealing with learning large amounts of information, one needs to be cautious in applying this method with older adults. Depending on the task, older people may have a difficult time absorbing the new information at the same time that they need to apply this knowledge to new task demands.

During this stage of learning, because a learner is still processing new information and there is a need to differentiate between the various facts, rules, and concepts, task performance may be unstable and characterized by slow improvement. Learners may even take steps backward in order to find an appropriate sequence of activities that lead to a successful outcome, or may need to rethink a term or concept or what constitutes a rule. However, at the conclusion of this stage of learning, learners are likely to have a basic understanding of the requirements for successful task performance, although they still may have gaps in knowledge that prevent smooth performance.

Generally, it is also a good strategy to try to induce some extra learning effort early in training by including some motivating exercises or those that require generalizations to different situations. Inducing extra effort may promote better retention of this information as well as the ability to apply information to different contexts. As an example of inducing extra effort, when training a person to perform a scheduling task (e.g., scheduling work parts in manufacturing or scheduling elective surgery schedules of physicians), the "trainer" (who could be a person or computer) can introduce early during training several unique types of conditions that the scheduler may face. For example, a number of machines that process work parts can simultaneously break down or several physicians scheduled to perform surgery can suddenly become ill.

Although the learner may not have accumulated the requisite skills to handle these situations smoothly, these types of problems can promote new avenues of thinking about the task elements and thus can facilitate the transfer of information to long-term memory, and the access of information from such memory. However, it is also important to help learners to build up sufficient confidence and to prevent them from becoming frustrated. We have found that older adults tend to experience greater frustration than their younger counterparts during learning of novel, complex work tasks.

Finally, it is recommended that "overlearning" of facts and procedures as well as basic task components, such as using a mouse or scrolling, be promoted. Overlearning will allow learners to acquire these tasks as more habit-based and allow them to depend less on working memory, so they will have more capacity to manage the difficult aspects of the task. This is especially important for older people, given that older adults have some reduced capacity for information processing.

The designers of training programs or instructions need to pay careful attention to issues related to legibility of information and the clarity of facts and concepts. If a reference is being made to a button on a device, the identification of this button and the legibility of any associated text or pictures should be clear, the specification of the purpose of this button should be unambiguous, and the distinction between this button and other buttons should be readily apparent. Neglecting these design-related considerations can disrupt the momentum of early learning and seriously impair the learning process.

Stage two

During the second stage of learning, people attain sufficient command of the background information, facts, and concepts to shape this information into "packets" of knowledge. The use of these packets allows people more efficient use of learned information and smoother performance of the task. An example of a type of packet is the "if–then rule." In the learning of calculator use, an example of an if–then rule would be "If I am in math mode, I can access a function that allows me to compute factorials."

In this stage of learning, integrating task components into larger components of knowledge can also be manifest as a refinement in strategies for performing the activity. For example, in the process of learning photography, a person may be able more successfully to capture objects at high speed by considering the effects of illumination in addition to shutter speed.

Stage three

Depending on what has to be learned, Stage three often requires the longest practice period. For tasks requiring a reasonable degree of skill, improvement in performance is much more gradual during this stage as compared to the previous stages and can occur over long sequences of repetitions. The rules developed in the previous stage become more habit-based, and people use less conscious thinking to decide how to perform an activity. During this stage, the procedures or strategies developed in the preceding stage are modified and fine-tuned to increase their reliability and efficiency. This fine-tuning can be accomplished in part because the person has refined the ability to discriminate between facts and concepts.

Similarly, the ability to select appropriate procedures in performing activities may require learners to identify connections between rules or between rules and facts. For example, the decision not to activate an auxiliary power supply can be inferred from linking the following rule and facts: if pump A is not working, either there is no electrical power to the pump or its switch is defective; pump D2 shares the same power supply as pump A; pump D2 is working.

Helping people to learn and remember

An important goal for training is to ensure that people will remember what they learned after periods during which the task is not performed. For example, it is critical that people who have diabetes and must use a blood glucose machine remember how to use the device after receiving training from their health care provider (see Chapter 9). For many to-be-learned tasks, retention of new skills and knowledge can be particularly challenging for older adults. In addition, they may use certain technologies, such as ATMs, computers, or VCRs, on a less frequent basis than might younger adults. Therefore, it is critical to ensure that the learning is not just for immediate mastery. It is also critical to provide the opportunity for "refresher training."

In general, the most important factor for ensuring that material is retained is to ensure a high degree of learning and to facilitate linkage of the new information with information that is already learned. An effective training strategy is to increase the amount of practice or to extend the amount of practice. Overtraining (i.e., training beyond performance mastery) can accomplish this goal. As discussed, another strategy is to inject difficult and varied problems into the training process to provide learners with opportunities for deeper levels of mental processing of the material. For some tasks, such as rule-based tasks that require careful discrimination of the relevant information before action, strengthening the degree of learning would require ensuring that learners are aware of the subtle differences in conditions that may exist and thus affect rule selection. For example, a person taking one type of medication for symptoms A, B, and C and another type of medication for symptoms A, B, and D needs to be able to discriminate between these two different patterns of symptoms. For some tasks, the use of a task analysis (see Chapter 3) can help to identify those aspects of the task that may require more practice.

Another technique that facilitates retention and improves the ability to apply the training material to situations outside the training environment is to use an instructional strategy that combines prescriptive information (i.e., knowledge of how to do things) with structural and functional information (i.e., knowledge that helps learners to understand the fundamental concepts that underlie how something works). An example of this approach would be a strategy that combines the basic principles behind a search engine and the organization of information on the World Wide Web with techniques for how to navigate the Web to access information.

We now turn to a number of training-specific issues related to helping people to learn and remember. These issues are especially important for older adults.

Organization of the training material

Training materials should be organized in a systematic way so as to maximize learning. For example, assume a person is being trained to perform a customer service task that requires responding to various types of customer queries by consulting the company's database. By grouping the information in the database into categories of interrelated items (e.g., company rules and procedures, product information, and customer information) and subcategories, the burden on memory processes can be reduced, resulting in more efficient retrieval of information in response to a query. During training, emphasis can be given to decomposing the query into relevant items of information that the trainee can then more readily link to the appropriate locations in the database. Providing links between items of information and corresponding categories of information through various strategies for organizing material is especially helpful for older adults, as it makes it easier for them to make appropriate distinctions and associations between various types of information.

Organizing to-be-learned information into related categories and concepts can also help to strengthen associations between the new information and information in long-term semantic memory. This enhances the likelihood that the new material will be available for use at a later time and is especially helpful for older adults. The benefits of organized training material also extend to perceptual-motor tasks whereby a learner is taught how to do something (e.g., how to control a motorized scooter). In these types of situations, organization can be accomplished through various cues, such as displays or voice prompts that guide trainees to appropriate actions.

Information should also be organized according to level of difficulty, proceeding from simple basic concepts to the more complex. For example, people need to learn basic mouse and window operations before they successfully search the Web for information. Similarly, we found that in teaching someone how to perform a data entry task associated with the trucking industry, it was important to provide training on how to discriminate between simple rules that govern where information is entered (e.g., "If the truck traveled only in the United States, mark field 1; if it also traveled in Canada, mark field 2") from more complex rules (e.g., "If there are no signs of having used excess fuel, leave column blank; otherwise, enter the amount of excess fuel used and location of its use").

Consistency of information

In presenting information, it is important to highlight consistencies that may exist between elements of the task and the response requirements. For example, in learning a software application, if the appearance of an icon requires one type of response during some of the learning trials and a different response in other trials, the icon is not consistently being linked to a response. Older adults are particularly susceptible to problems arising from lack of consistency, owing to changes in attention and perceptual abilities. For example, a common problem with cellular telephones is a lack of consistency in the location and operation of the function buttons (e.g., "talk," "end call"). This not only may create difficulty when a user is attempting to learn to use the phone but can result in frustration. When response requirements are consistent only within certain situations, people can learn how to respond in that particular situation. Training needs to emphasize "situation-specific" consistency. A thorough task analysis prior to developing a training program or strategy is essential to determining how to teach the consistencies of the task.

An important benefit of incorporating consistency into training is that it can lead to "automatic" responding. Automatic responding is relatively fast and does not rely much on attention. This can be especially important if stressful or emergency conditions arise, such as interacting with a medical device while ill or responding to alarms in a work situation.

Distributed versus massed practice and temporal spacing considerations

Two contrasting training approaches are to provide multiple exposures of the material over time (distributed practice) and to concentrate the exposure of the material in a single training session (massed practice). Generally, the available data suggest that for a wide variety of tasks, especially for those that are complex, distributed practice is more advantageous than massed practice with respect to skill acquisition and retention.

It is also important to consider the duration of the training sessions and the interval between successive sessions. For most learning situations, very short rest breaks between study sessions (e.g., less than 5–10 minutes) should be avoided, although the actual times would depend on the duration and complexity of the training sessions. Thus, trainers must exercise some degree of judgment in establishing the length of a rest break. Also, in general, short training sessions between rest breaks are superior to long sessions. If the training sessions are long (e.g., > 30–45 minutes), rest breaks are required. Rest breaks have numerous benefits. One benefit is that people are less likely to have to process and retain too much information in a given period. Rest breaks obviously help to prevent problems with fatigue. We have found that in training older adults for complex work tasks, rest breaks of approximately 15 minutes between study sessions provide older adults with a greater opportunity to think about and review the learned material and help to minimize interference with subsequently introduced material. However, in these situations, "refresher training" may have to be provided to ensure adequate retention prior to the introduction of new study material.

For tasks that are not complex, such as learning to use an ATM, and for tasks that are largely perceptual-motor activities, such as learning to operate a VCR, long intervals between study sessions may adversely affect older learners. Older people are often more anxious about their performance and thus can gain much needed confidence by perceiving an attainment of mastery within a reasonable period of time. For these types of tasks, the availability of memory aids for use after the training (e.g., a laminated instructional card) can be useful. These types of aids provide knowledge in the world to support performance rather than requiring the users to have the knowledge in their minds.

Mental models and analogies

As noted earlier, training techniques that facilitate thinking about familiar concepts and relationships can also strengthen associations in memory and thereby improve retention. A schema is an organization of knowledge that we have in memory about a particular concept or topic. When this knowledge concerns devices and systems, we often refer to this organized knowledge as a *mental model*. People may use mental models to determine how to use a word-processing system to format a document in a particular way; how to program a VCR to record a number of different future programs; how to dismantle a hazardous device; or when to take a boat out into the ocean. In forming their mental models, people often use analogies. The benefit of providing analogies during training is that they allow the learner to use existing and familiar knowledge better to understand components of the task being learned.

Training processes that help learners to build mental models or to identify analogies can facilitate retention of what has been learned. Having a mental model or an available analogy can help people to recall a fact or other information or to recognize information that is being presented. When older adults are being trained to use new, unfamiliar technologies, helping them to construct a mental model or to identify a familiar analogy can lessen the anxiety associated with learning new concepts and operations. The basic idea is to help them to place these new concepts and operations into a familiar context. As an illustration of this principle, a good strategy to use in teaching novice computer users to manage computer databases is to discuss the task as if it were removing a paper folder from a file cabinet drawer. To create proper analogies, the instructional designer must have a thorough understanding of the to-be-trained group of people. The trainer must also recognize that not all analogies are understood by all people.

Part-task versus whole-task training

Generally, breaking a task into manageable learning units is a desirable strategy. In view of the recommendations regarding time-duration effects, it would also appear that such part-task training would be especially beneficial to older people. However, the superiority of part-task training to whole-task training (whereby training is provided on the entire task at once) depends on how the task is broken down and on the nature of the task. For all tasks except those that are very simple, some degree of part-task training is recommended. If a task is readily amenable to part-task training, this training strategy will be particularly suitable for older people because it imposes fewer demands on working memory and provides an opportunity for learners to master each task component.

For example, consider an older worker who is being trained to perform various operations related to the production of a pharmaceutical skin-care product. The operations required include connecting various pipes and hoses, measuring the amounts of the different ingredients, timely activation of the appropriate pumps that will discharge the ingredients into a vessel, and the monitoring of the blending of these ingredients into the final product. This overall operation lends itself to partitioning into component operations, making it advantageous to train these task components separately. The advantage of this training strategy becomes more apparent if the task components are not equal in difficulty; strategies can then be adopted that allow allocation of more training time to the more difficult task components.

In this example, the task components are performed sequentially. In other cases, as in training a worker to manage an automated manufacturing system that processes many different types of products, many of the component tasks are performed in a more simultaneous rather than sequential manner. Examples of these tasks are scheduling, control of congestion, making decisions on reprocessing parts, and handling machine breakdowns. In either case, it is important that learners understand the relationships among task components and the Gestalt of the task as a whole. Training programs can achieve this objective through the use of a task analysis that emphasizes the nature of the interrelationships among the component tasks. Such a task analysis should highlight the existence of constraints in the sequence (e.g., task D must immediately follow task B) and in the timing (e.g., 10 minutes should pass before ingredient A is added) of the component tasks. The task analysis should also identify contingencies (e.g., if more than five parts are waiting to be processed at machine C, try rescheduling the parts that are toward the end of machine C's queue).

Adaptive training

When training is flexible rather than fixed and is provided in accordance with the progress and needs of a particular person, the training is called *adaptive*. This type of training is generally feasible only when applied to individuals rather than groups of people and has advantages for older people who may have a wide variety of capabilities and limitations that influence learning.

The key attribute of any adaptive training program is that it can make inferences about a learner's knowledge and can then adapt the training process to accommodate best the trainee's current learning needs. Whether people or computers carry out adaptive training, a thorough task analysis is necessary for anticipating problems that people with different degrees of knowledge and skill may have at each step of training. Strategies then have to be developed for handling these problems. Thus, if a person is having trouble learning how to use a financial-management software application, the training program may alter its route and choose to focus on finance terminology or on particular software operations, depending on the problem the person appears to be having.

Owing to the flexibility that can be incorporated into software, computer-based instruction is a popular means for implementing adaptive training, although there is no evidence that use of computers per se will automatically improve the learning process. The process of making inferences that underlie adaptive training can be seen on much of today's personal computer software. In this sense, the concept of adaptive training bears some relationship to intelligent tutoring or help systems.

Feedback

Providing feedback helps learners to correct mistakes and reinforces procedures and concepts. It is also important with respect to motivation. Thus, both positive and negative (albeit constructive) feedback should be provided during training. Providing feedback during training minimizes repeated errors, particularly important for older adults for whom unlearning is particularly difficult.

Feedback can vary in a number of ways. It can be immediate (e.g., telling learners as they are ready to select a menu that the selection is incorrect) or delayed (e.g., telling learners after successful or unsuccessful completion of a task which actions should have been taken). It can also be given on a frequent basis (after each step) or an infrequent basis (at the end of the training session). Depending on the training situation, it can also be provided through face-to-face communication or through synthetic voice or text displayed on computer screens or printed instructional manuals. Another factor to consider is whether the feedback provided should be extensive (e.g., providing the conceptual basis for why an action is wrong) or be relatively brief (e.g., providing the number of wrong answers). Thus, there are many options to consider in providing feedback, and the choice depends on the training situation. Group training may preclude providing extensive or frequent feedback to each individual in a group unless sufficient time and training personnel are available.

With older adults, it is essential that adequate feedback is provided during the early stages of learning, and this feedback should be immediate. Providing immediate feedback is important in teaching basic computer skills, such as scrolling, double-clicking on a mouse, or dragging. Such feedback ensures that these actions are understood and do not impede the learner's ability to perform other computer applications, such as word processing or searching the World Wide Web.

Older adults generally will benefit greatly from face-to-face feedback. This form of feedback is especially useful during the early stages of learning when anxiety levels may be high. If training is being carried out through an instructional medium, such as a computer or a printed manual, the designer of this training material needs to anticipate incorrect or inappropriate actions that an untrained user may attempt, and to provide feedback that can address those actions. Suppose it is anticipated that a user may attempt to obtain a reading from a medical device when the device is not set to the operational mode that produces this reading. The user then should be made aware of this discrepancy. This can be done through a visual display or through printed material organized into an appropriate section in an instructional manual (e.g., a section that reads "Why you may not be able to get the machine to display a blood pressure reading").

Although feedback is essential, especially during the early stages of learning, providing too much feedback is not a good idea. Excessive feedback disrupts the learning process because it can cause people to focus on the wrong things or it can overload working memory. Thus, there is often a fine line between providing repeated or excessive feedback that can be overtaxing and providing insufficient feedback that can hamper the learning process.

Several of the suggested recommendations were adopted in a study that we conducted involving training older adults to perform a simulated telecommuting work task. The older adults were trained in groups of six at individual computer workstations, and the trainer had a facilitator available to address technical issues regarding use of the computer, mouse, scrolling, and other basic computer operations early on in training. The study material was subdivided into relatively small modules, each containing topical information that was considered to be well within the working memory capabilities of the learners. Depending on the size and content of the module, questions regarding the material and feedback were provided after key subsections and also at the conclusion of the module, to gauge adequate understanding of the concepts. Demonstrations using test e-mails then followed to provide a context for applying those concepts. After the conclusion of all the modules, each trainee had the opportunity to open and process six e-mails from fictitious customers. They received explicit feedback concerning their performance on each of these e-mails through screen displays. The trainer then went through each e-mail in a review session that allowed trainees to examine what they did incorrectly and to question why their actions were inappropriate. This approach established a strong (accelerated) learning curve. Subtle issues that prevented performance from reaching very high levels could then be addressed during the later stages of learning using feedback directed at further explication of the study material. Overall, the strategy was found to be effective in teaching people of all ages this telecommuting task.

The use of simulation in training

Learning many activities is often facilitated by the use of task simulations. These simulations can take the form of sophisticated physical devices equipped with computer and electromechanical controls (e.g., an airplane cockpit or nuclear control room simulator) that attempt to capture the feel and conditions of the real world, or they can embody less sophisticated mock-ups (e.g., the use of lifelike mannequins for training on CPR). Many tasks can be simulated using computer interfaces (e.g., software that emulates an ATM). Most evidence suggests that even if the simulated task environment possesses low fidelity, it can still be a very effective training device as long as it captures the essential relationships between the task attributes or conditions and the appropriate responses to those conditions. The development of successful task simulations is highly reliant on detailed task analyses.

One reason why simulation is used in training is that trainees tend to learn best if they are not afraid of making errors during training, especially during the early stages of learning and on tasks requiring complex decision making. This factor makes simulation particularly suitable for training older adults, especially on unfamiliar tasks. Older people tend to be more anxious about their performance and aware of their limitations. The ability to experiment with task performance without exacting any penalty with regard to the risk of personal injury, damage to property, or economic loss can provide older adults with the confidence needed to overcome initial fears. This may, in turn, make training more cost-effective. Simulation often also provides learners with better control over the pace of practice and the opportunity to be exposed to a greater array of examples. Overall, the tremendous flexibility in the control of learning and performance assessment afforded by simulation makes it an extremely effective approach to training, especially for older adults.

Instructional media

With advances in computer technology, the variety of instructional media available has grown, making it difficult to decide which media should be used for a training program. Currently available media include such printed materials as charts, graphs, maps, checklists, manuals; audio tapes and CDs; static projections in the form of slides, overheads, and photographs; video-motion, such as television and video; and computer-based training using expert systems and other forms of intelligent instruction.

Bounding these instructional media are traditional lectures and simulations (that can include role playing as well as high-fidelity physical simulators). Complicating the picture is the fact that training can be carried out using many different combinations of these media.

Research on instructional media has resulted in many unanswered questions, and there is some evidence that innovative technologies provide only very slight gains over traditional instructional strategies. Many experts in this area have abandoned the notion of discovering the best instructional delivery medium and have recognized that the instructional strategy depends on a number of factors, including the task, the characteristics of the trainee population, feasibility, and resources. In fact, there is a widespread belief that all media can deliver effective or ineffective instruction and that it is the training methods embedded within the instructional medium that have the most pronounced effect on learning.

There is some evidence to suggest that specific media work best for particular types of training situations. Some media may provide a better understanding of which buttons to press in using a product (e.g., by watching a video), whereas other media (such as a direct demonstration) may make it easier to visualize how to hold a surgical instrument relative to the position of the patient. Media that present motion have been found to be more effective for activities that involve human movements in different directions that may not be easily or accurately described in words. Likewise, the use of animation (e.g., in depicting how the values on a medical device change) may be more effective than static illustrations or photographs for situations wherein change or movement over time is critical for understanding how something functions.

However, there are situations wherein the ability for the media to enhance greatly the details of presentation, such as an aircraft engine or a body organ, may not result in improved learning; appropriately detailed line drawings may actually be more effective for more advanced learners. This point illustrates the potential concerns with delivery of instruction to older adults through various instructional media. It is important not to overload the learner with too much or too detailed information. For example, multimedia instructional formats may be too distracting or provide too much information. Therefore, if this type of instructional medium is used, it has to be carefully designed to ensure that it provides the best combination of clarity or insight into how to perform the task. As discussed in Chapter 4, careful attention also has to be given to the presentation of visual and auditory information to accommodate age-related changes in sensory and perceptual abilities.

Guidelines for the design of training and instructional programs

The following section presents design guidelines that summarize the information presented in this chapter. These guidelines are important to consider in designing training programs for older adults. As noted, most of these guidelines would also benefit learners of all ages.

- Allow extra time for training older adults (one and a half to two times the training time expected for young adults).
- Ensure that help is available and easy to access and that a person is acquainted with all available sources of help. For computer-based training, ensure that labels for help functions are intuitive

to the user population, and provide a tutorial for use of the help system.

- Ensure a training environment that allows people to focus on the training materials (e.g., by minimizing such distractions as background noise, other worker activities, too many instructors, and the use of multimedia).

- Ensure optimal organization of training materials and provide a structure with clear identifiers, headings, and subheadings. Also ensure that the reading level of all instructions and manuals matches the abilities of the user population; that illustrations that are provided present specific examples of the learning materials; and that the requirement for inferences is minimized.

- Point out consistencies that may exist between elements of the task and the response requirements.

- Match instructional technique and medium to the type of material that is being presented. For example, "how to" information should be presented in a procedural step-by-step format, whereas spatial tasks are best trained using a visual medium.

- Allow learners to make errors, when safe, but provide immediate feedback regarding how to correct mistakes, especially during the early stages of training.

- When needed, reduce training demands by using part-task training techniques to provide practice on task components. In doing so, proceed from simple to more complex aspects of the task, and ensure reliable performance on basic components before moving on to higher-order components.

- Provide opportunities for learners to be actively involved in the learning process by creating an engaging learning environment that captures the attention of learners. For example, provide opportunities for learners to manipulate a device or be involved in the solving of practice problems and inject some reasonably challenging exercises.

- Minimize demands on working memory: provide cues and aids, do not overload learners with too much information, provide feedback in a timely manner, and do not require learners to make complex inferences and fill in gaps of missing information.

- Ensure that learners have adequate understanding of basic concepts and prerequisite knowledge to benefit from the training program. Also, provide remedial training as needed (e.g., in training individuals to use a Web browser, they must be able to use a mouse and understand basic concepts of windowing interfaces).

- Capitalize on users' preexisting knowledge base (e.g., use analogies or metaphors to link new information to familiar concepts and, wherever possible, avoid technical jargon). Make sure the analogies are appropriate by thoroughly understanding the diversity that may exist in the to-be-trained audience.
- If training sessions are relatively long (e.g., > 30–45 minutes), rest breaks of at least 5 to 10 minutes should be provided. If training sessions extend across a day or over several days, provide refresher training prior to the introduction of new study material.

Using these guidelines to design a training program

This chapter summarizes the importance of providing training and instruction and summarizes the issues that need to be considered in developing training programs for older adults. However, as discussed in Chapter 3, the development of a training program should be a systematic and iterative process. At a minimum, the instructional design process must consider the task, the characteristics of the trainee population, time, cost, and available resources.

Recommended readings

Brock, J.F. Computer-based instruction. In G. Salvendy (ed.), *Handbook of Human Factors and Ergonomics*. 2nd edn New York: John Wiley, 1997:578–93.

Czaja, S.J. Aging and the acquisition of computer skills. In W.A. Rogers, A.D. Fisk, and N. Walker (eds), *Aging and Skilled Performance: Advances in Theory and Applications*. Mahwah, NJ: Lawrence Erlbaum, 1996:201–20.

Gagné, R., Briggs, L.J., and Wager, W.W. *Principles of Instructional Design*. New York: Holt, Rinehart, and Winston, 1989.

Rogers, W.A., Campbell, R.H., and Pak, R. A systems approach for training older adults to use technology. In N. Charness, D.C. Park, and B.A. Sabel (eds), *Communication, Technology and Aging*. New York: Springer Publishing Company, 2001:187–208.

Swezey, R.W., and Llaneras, R.E. Models in training and instruction. In G. Salvendy (ed.), *Handbook of Human Factors and Ergonomics*. 2nd edn New York: John Wiley, 1997:514–77.

Wickens, C.D., Gordon, S.E., and Liu, Y. *An Introduction to Human Factors Engineering*. New York: Longman, 1998:553–92.

Chapter 6

Design of input and output devices

An input device is a mechanism to communicate intention or action to a technology system. The button is a classic input device. Pushing buttons to get a receipt from an automated gasoline pump, to fix time and temperature settings for a microwave oven, to dial a number on a telephony device, are all examples of actions taken with input devices. So, too, is speaking into a microphone system to order a meal from a drive-through restaurant or to respond to an automated voice response system with a menu item (e.g., in response to the prompt to "Press or say one").

An output device is a mechanism that communicates with a user. Examples would be microwave ovens that signal the end of a cooking interval with sound and a visual message; a computer screen that signals states of a program or operating system with visual and auditory messages; and a global positioning system that is used for navigation and provides voiced instructions to turn left or right at intersections.

In some cases, the input and output devices are the same (e.g., touch screens on hand-held computers that permit stylus input as well as visual and auditory output). Irrespective of the type of device, people must interact with it via their sensory-perceptual and psycho-motor systems. As Chapters 2 and 4 indicate, aging processes affect sensory and perceptual systems as early as middle age, so the benefits of good design can be seen soonest with input–output aspects. For example, loss of ability to focus on near objects becomes evident in the early to mid-forties, making it difficult to focus on text that is displayed in the upper part of the visual field. For those who wear bifocals (reading distance lenses in the bottom part of the spectacles and distance vision in the upper part), viewing a computer screen in the upper visual field requires craning the head up to read through the lower bifocal. Special bifocal lenses that have middle distance in the upper half and near distance in the lower half can help to circumvent this problem, but many people will not have such lenses (or may not be wearing them when they interact with a device).

Because of the high prevalence of arthritis after age 65, such simple actions as using a keyboard or a mouse can become uncomfortable, and this can be exacerbated by requiring acquisition of very small targets, such as an icon on a computer screen via a pointing device. On average, 50 percent of men and 60 percent of women older than 75 report having arthritis, and the hands are a very likely location for the disease. One implication is that completion of keystrokes may be uncertain for those pressing keys gingerly; hence, it might be useful to supplement the usual minimal tactile feedback of a key press with an auditory signal.

A tremendous amount of work has been conducted on issues related to the design of both input and output devices (e.g., *Handbook of Human Factors and Ergonomics*, 1997; *Handbook of Human–Computer Interaction*, 1997; *The Human–Computer Interaction Handbook*, 2003), and standards have also been developed specifically for workstation equipment (BSR/HFES 100, 2002). The intent of this chapter is to highlight those issues particularly relevant to older adults.

Issues in aging and the design of input devices

A number of issues should be considered in designing input devices that will be used by older adults. As Chapter 2 indicates, dexterity and strength change negatively with age. Speed declines. Vision and hearing may be impaired. Speech patterns become more variable. All these age-related changes make interactions with the environment—formerly very easy and reliable in young adulthood—increasingly problematic in old age. Older adults are also more likely to suffer cognitive decline that may make them slower and more error-prone in mapping their actions to devices. A good example is the computer mouse that requires adjusting to its gain (greater speed and acceleration of the screen cursor position than of the mouse). Older adults are more likely to have problems in controlling fine motor movements because of arthritis or tremor. Designers ought to consider providing alternative ways to navigate with input devices.

Minimizing steps

Input devices enable people to select a broad range of actions via a sequence of command activities. Except in the rare case of "chording" devices that allow multiple simultaneous inputs, commands are usually organized serially. On a television remote control, for instance, the user first powers up the television set by depressing a power button, then uses other sequences of key presses to select a channel, adjust sound volume, and so on. An error anywhere in the sequence can block the user's goal (e.g., omitting the first step makes successive button presses futile). The difficult tradeoff for designers is to minimize both the number of steps in the procedure (e.g., button selections and presses) and the number of controls (e.g., buttons). One could imagine a remote control with a large set of buttons that both turn on the television and select each channel, but that remote would be huge and unwieldy, not to mention expensive.

There is a heuristic urged on those designing procedures: KISS (keep it simple, stupid). Here we propose MS (minimize steps). The importance of minimizing the steps needed to achieve the user's goal can be seen in the following example. Assume that there is a constant probability of error on a given step (in reality, some steps may be more error-prone than others) and that every step must be carried out correctly for the correct action to occur. If the steps must be carried out serially and even if the reliability of the user in carrying out any step is high, the probability of successfully completing the entire procedure drops off sharply with the number of steps: $p\,(\text{success}) = (1 - p\,(\text{failure}))^{N\,(\text{steps})}$. An example is shown in Figure 6.1. The three lines represent the cases for failure rates per step of 1 in 100, 1 in 20, and 1 in 10.

It does not take many steps before there is only a 50-50 chance of success for the procedure as a whole, even with relatively low failure rates: 7 steps for a 1-in-10 chance of failure per step and 13 steps for a 1-in-20 chance of failure per step. It pays to minimize the number of steps, even in situations in which the cost of failure is only the time to redo the procedure (e.g., selecting a television channel to watch) rather than the potential life-or-death situation entailed by carrying out a medical procedure correctly on the first try. Lest one think that common procedures have relatively few steps, a task analysis of a "simple" procedure ("as easy as one, two, three") for using a blood glucometer actually entailed more than 50 steps!

Another common abuse of the MS principle is requiring the entry of long character strings for software authentication codes. For

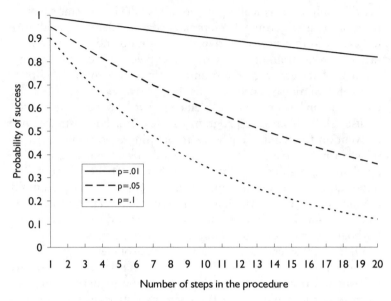

Figure 6.1   Probability of success as a function of the number of steps and the failure rate for each step.

example, instead of the bad practice of requiring a 16-character authentication code, a software designer could select a short 5-character authentication code, such as aw4p7, which is difficult to guess given 36^5 possibilities for values for letters and digits, but easy to read, remember, and enter.

Do older adults have higher error rates on average than younger adults? The literature on error rates for simple decisions (yes versus no reaction time) suggests that older adults are somewhat more likely to emphasize accuracy than speed in their choices in contrast to younger adults, who are more likely to do the opposite. However, when it comes to tasks with significant memory demands, older adults are more likely than younger ones to commit errors in reproducing a long sequence (e.g., digits in a digit span task). Hence, minimizing the number of steps in a procedure is particularly helpful to older users.

Consistency

Consistency in layout of control elements is usually a very important design goal. The success of the rental car industry has depended heavily on having critical input and output devices in standardized positions (e.g., steering wheel, brake pedal, accelerator pedal, turn indicator) so that licensed drivers can walk up and use the vehicle safely with no training. Control elements less critical (to driving) often vary widely (e.g., radio and music device controls, climate controls). Similarly, software display interfaces strive to ensure consistency in layout of control elements in order to facilitate transfer of user skills from one program to the next. For instance, one convention for windowing software is using a pointing device to select a window, making it the active window, and signaling selection by changing highlighting for the top border of the window. The importance of consistent mapping (e.g., of keys to actions) is painfully obvious to those of us who have ventured into foreign Internet cafés to check e-mail and have been forced to use a different country's keyboard to carry out commands and to input text.

Older adults tend to rely more on environmental support for memory processes. That is, they rely on external cues to retrieve information from memory (such as the action sequence they need to carry out to achieve a goal). Given their stronger reliance on a consistent mapping between stimuli and learned responses, they suffer more interference when conventions are ignored, so in this group of users, consistency in the layout of controls is even more important than usual. If the design flouts the user's expectations, older adults, who generally learn more slowly, will likely be inconvenienced for a longer period.

Another concern with older adult users is that the software interface conventions for controls are often less familiar, particularly for novice older users. Some have argued that older cohort members learned conventions for controls at a time when there was almost always a one-to-one mapping between control type and outcome. Today, software interface controls exhibit many-to-one mappings that often depend on the mode (state) of the display. For instance, the tab key maintains a typewriter-like function of shifting the cursor (and any subsequent text strings) to the right when used within the active part of a text entry window for word processing. However, the tab key moves highlighting to different menu items in a top-to-bottom direction when used within drop-down menus and jumps to the top element after highlighting the bottom item of the menu column. Also, the tab key does not function at all for the Windows Start menu (although cursor keys do).

In general, it is good practice to design for expectations or what have been called "affordances" (e.g., visual aspects that suggest function, such as the way a handle affords grasping and pulling) and to check whether expectations vary across generations (age groups). As an example, standard two-position light switches provide affordances in terms of up and down positions for the physical switch. The idea of "toggling" between settings brings to mind two positions (e.g., on–off). When this convention is violated, it takes significant problem solving to understand the new functions. For instance, holding down the shift key and striking the F3 key in the popular Microsoft Word program toggles between three states in terms of actions on text. (Depending on the state of the text, a further complication—text changes such that the first character in a string becomes upper case and the rest remain lower case—all characters in the string become upper case, or all become lower case.)

Types of input devices

Input devices have many uses, from authentication (e.g., traditional door keys) to action (e.g., keyboard keys). There are many input devices available commercially, some highly specialized for a particular task. We discuss the main categories. As is usual, no single input device is ideal for every type of task; each has advantages and disadvantages. We evaluate these devices stressing the perspective of older users.

There are many ways to classify input devices and, to simplify the discussion, we view them from a functional perspective. Table 6.1 provides an overview of input devices (important attributes for the device are given in columns). Although we have older adults as the target users in mind, principles should hold for younger adults.

A good example of a ubiquitous input device is a remote control unit that communicates with a television or other multimedia device. According to a *New York Times* article a few years ago, there are more remote control devices than people in the United States! The devices shown in Figure 6.2 illustrate some good and bad design principles for remote controls.

Guidelines for the design of input devices

• Select good default values or develop profiles that could be selected on the basis of different age groups (children, adults,

Table 6.1 Input device types and characteristics

Input device category	Cost	Training required	Calibration required	Pointing or moving		Tracking		Text and data entry		Comments
				Speed / Precision		Speed / Precision		Speed / Precision		
Positioning (indirect)	Low	Medium	Low	Medium / High		Medium / Medium		Low / Medium with software keyboard		Good for experienced users
Mouse (simple one-button, two-button, optical, wireless)	Low–medium	Medium	Low	High / High		Medium / Low		Low / Medium		Avoid interfaces requiring double-click selection of targets
Track ball	Medium	Medium	Low	Medium / High		Medium / Medium		Low / Medium		Can be used in double-click selection fairly easily
Joystick and keyboard joystick ("trackpoint")	Low–medium	Medium	Low	Medium / High		High / High		Low / Medium		Best for tracking tasks
Rotary encoder	Medium	Medium	Low	Medium / Low		N/A		N/A		Good for precision tasks or repetitive movements; less variable for older adults relative to a touch screen

Graphics/Touch tablet (relative or absolute setting)	Medium	Medium	Low	Medium / High	High / High	Low / Medium	In absolute mode, it resembles a direct positioning device; high risk of accidental input
Positioning (direct)	High	Low	Medium	High / Medium	High / High	Low / High with software keyboard	Uses natural responses (gazing, pointing to target); higher workload; can obscure display during movement
Touch screen (resistive, capacitive, infrared, piezo-electric, crosswire)	High	Low	Medium	High / Medium	High / High	Low / High	
Light pen	High	Low	Medium	High / High	High / High	Low / High	Usable with CRTs only
Eye-movement control	Very high	Medium	High	Medium / Medium	High / Low	Low / Medium	Need highly trained personnel to use this system; targets should exceed one degree of visual angle

(continued ....)

Table 6.1 (continued)

Input device category	Cost	Training required	Calibration required	Pointing or moving	Tracking	Text and data entry	Comments
				Speed / Precision	Speed / Precision	Speed / Precision	
Data entry	Medium	High	Low	Low / Low	Low / Low	High / High	
Keyboard / Keypad	Low	High–medium	Low	Arrow / page keys for navigation: Low / Medium	Arrow keys: Low / Low	High / High	Problematic for those with dexterity impairments; try to minimize the force required to trigger keys; provide adjustable auditory feedback, especially for membrane keypads.
Software keyboard	Low	Medium	Low	N/A	N/A	Low / Medium	Prefer QWERTY layout of keys for novice users; use 10 × 14 mm keys and interkey spacing of 19 mm center to center; provide feedback for "presses"

Hand-writing recognition	Low	Medium	Low	N/A	N/A	Low / Medium	Accuracy problematic without extensive training
Speech recognition	Medium	Medium	Medium	Low / Medium	N/A	High / Medium	Best in noise-free settings; difficult for correction of errors

N/A = the category is not applicable. Low, medium, high = relative effort, precision, etc. required by the device.

Note:

Speed and precision are considered for the case of already trained individuals. For use of multiple devices (e.g., pointing and data entry), consider the problems of "homing": moving the hands to the home row key position from the pointing device (e.g., homing is a more severe problem for a touch screen or light pen than for a mouse or track ball).

Poor contrast for lettering on surface (black on gray); small, crowded number buttons; good contrast between black buttons and white text; good example of using redundant cues (button size and shape).

Good contrast for lettering (white on black); larger buttons that are well spaced; bad contrast between gray buttons and black background; good choice of contrast color and shape variation for step control buttons.

Figure 6.2   Two remote control devices that have good and bad design features.

seniors). Do not assume that flexible interfaces will result in optimal choice of parameters by users.

- Match the input device with the task demands. For example, using Table 6.1 as a starting point, prefer track ball to mouse for novices if the interface requires double-clicking; consider default interfaces that do not require double-clicking. Prefer direct (light pen, touch screen) to indirect (mouse, track ball, joystick) positioning devices for pure pointing and clicking tasks, particularly when the input device is not large (e.g., hand-held computer). Prefer indirect devices if users are experienced and the task requires combined keyboard entry and device use; if the extent of movement for a direct device is large (e.g., 19-inch monitor); or if the task requires precise selection. Prefer speech recognition control and input when individuals are very restricted in manual dexterity and the ambient noise level in the environment is low. Prefer cathode ray terminals to back-lit liquid crystal displays (LCDs) when precise color matching is needed.
- For keypad input, use large keys with clear markings (adequate contrast for text or symbol to background) and appropriate interkey spacing (see Figure 6.1).
- Provide for the possibility of both tactile and auditory feedback with keypads. (This situation occurs with many microwave ovens that emit beeps on key press and can be coded in software for computer keyboards).
- Permit alternatives for navigation with a visual cursor (such as arrow key movement) for those with moderate tremor.

Issues in aging and the design of output devices

Our focus here is on specific issues related to a given device and to visual and auditory displays. Other relevant issues, such as the information organization and format, are discussed in Chapters 4 (Improving Perception of Information) and 7 (Interface Design).

For the most part, output to users (feedback) comes via visual and auditory sources. For visual output displays, the concerns raised in earlier chapters (particularly Chapter 4) hold here. It is highly worthwhile to investigate the task environment in which a device is going to be used. Visual display screens are common in most electronic devices, appearing in everything from phone devices (e.g., to provide caller ID information) to electronic thermometers, to microwave ovens. A variety of display elements are used in these devices, from passive LCDs to light-emitting diodes. Lighting levels for reading passive LCDs (non-back-lit) are often adequate in office environments but rarely so in home environments. Consider using emissive displays (e.g., fluorescing) in such environments. Consider also the angle from which a display must be read in choosing the type of display. Although outdoor environments provide high ambient light levels, they are also likely to include glare sources. Shielding displays in outdoor environments is important. A good example would be shielding the passive LCD on a gasoline self-service pump or putting an ATM display indoors in a booth (which also helps users in poor weather conditions, such as rain or snow). Unfortunately, as those who drive automobiles in the direction of the rising or setting sun recognize, it is almost impossible to ensure optimal viewing conditions at all times of day even for interior displays.

Guidelines for characteristics of output devices

- Select an output device with the higher contrast between characters and background. For example, in choosing between an LCD and a cathode ray terminal, an important consideration is to select on the basis of the best contrast ratio for the ambient light conditions.
- Ensure that the size of the text to be read is 0.6 degrees of visual angle or greater on the display; for a four-letter word, this approximates the width of your thumb at arm's length.
- Maintain visual output screens adequately shielded from glare.
- Provide an adaptive (adjustable) display when feasible and provide instruction to users about how to change screen resolution.

  - Consider advising older users to set the resolution of their computer display to 800 × 600 pixels or lower (640 × 480 pixels) to enhance access to small icons typical in today's software interfaces.
  - Use built-in controls (e.g., Microsoft Windows accessibility functions available through control panel settings) or special purpose software.
  - Permit adjustability of output intensity and frequency of sounds.

- For important visual warning messages, repetitively flash information rather than having it appear without interruption. However, make sure that the flashing is not so fast as to impede reading of the message.
- Prefer tactile output devices for simple signaling (e.g., using moderate to high-frequency vibration, approximately 250 Hz) when auditory and visual output would be difficult to process (noisy environments, glare situations) or would be disruptive to performance by users or nearby personnel.
- For important auditory warnings, select output (e.g., speaker) systems that emit sounds in the 500 to 1,000 Hz frequency range and repeat the message until acknowledged.

Using these guidelines for effective design of input and output devices

In selecting input devices, designers must consider the type of activity required by the interface (software controls; see Chapter 7). If people are performing only point-and-click operations (e.g., database information retrieval within a graphic user interface), choose direct positioning devices. In mixed pointing and data entry tasks or for precision tasks, choose indirect positioning devices.

For output devices, designers must both "know the user" and "know the user's environment," given the differences in features of these environments, particularly for lighting. Light levels in homes for reading materials are typically in the 30 cd/m² range, compared to 100 cd/m² found in offices. Ensuring good contrast for output sources becomes critical in these environments. Consider substituting active or fluorescing for passive LCDs or provide back-lighting for LCDs when they are intended for home use. For portable devices that will be used both indoors and outdoors, consider transreflective displays, such as those found on second-generation palm-top computing devices and personal digital assistants.

No device is optimal for every task and every user; the best strategy is to provide guided choices to consumers. It is particularly important to test both input and output devices with the targeted population of users, following the guidelines in Chapter 3.

Recommended readings

BSR/HFES100 Human Factors Engineering of Computer Workstations. Santa Monica: Human Factors and Ergonomics Society, 2002.

Helander, M.V., Landauer, T.K., and Prabhu, P.V. (eds), *Handbook of Human-Computer Interaction*. 2nd edn. Amsterdam: Elsevier, 1997.

Jacko, J.A., and Sears, A. *The Human-Computer Interaction Handbook: Fundamentals, Evolving Technologies, and Emerging Applications*. Mahwah, NJ: Erlbaum, 2003.

Salvendy, G. *Handbook of Human Factors and Ergonomics*. 2nd edn. New York: John Wiley & Sons, 1997.

Steenbekkers, L.P.A., and van Beijsterveldt, C.E.M. *Design-Relevant Characteristics of Aging Users*. Delft, The Netherlands: Delft University Press, 1998.

Chapter 7

Interface design

An interface, by definition, is the point at which two systems meet and communicate with each other. The focus of this chapter is on the human–computer interface issues related to menu design, display layout, system navigation, and design of help systems. Input devices are reviewed in depth in Chapter 6, and issues of training and instruction are the focus of Chapter 5. Computer use in the workplace is discussed in Chapter 8, and in the context of health care in Chapter 9.

Human–computer interfaces abound in our environment. Table 7.1 outlines the numerous activities for which individuals must interface with a computer system of some form to perform a multitude of tasks. This list could continue for several pages and does not even include common computer tasks, such as searching the World Wide Web, data entry, word processing, and using e-mail.

Table 7.1 Examples of tasks involving interfaces

Depositing a check into an automatic teller machine
Setting a videocassette recorder to record a program
Checking into an airport kiosk to obtain a boarding pass
Organizing appointments on a personal data assistant
Entering a security code for an alarm system in a vehicle or home
Recording a message on an answering machine
Programming a microwave to reheat leftovers
Adding names and telephone numbers to a cell-phone list
Changing the time displayed on a car radio
Taking a reading from a blood pressure monitor
Finding a book using the online library catalog
Selecting an option from a recorded telephone menu system
Taking a picture with a digital camera

Clearly, most people must be capable of interacting with a variety of display layouts, menu systems, and navigational aids. Further, most current computer users are not highly trained computer specialists: they are young and old, educated and less educated, experienced and novice. How can we ensure that such systems are usable by the range of users who will interact with them? Are there specific design requirements for systems to be used by older adults? These questions are the focus of this chapter.

Interface design issues

Interface design issues are relevant to a broad range of computer technologies. Computers are found in nearly every aspect of our lives, from the marketplace to the workplace, in the home and in the car, in the context of health care and in leisure activities. Computer systems may be differentiated according to whether they are intended to be used by virtually anyone who comes across the system (e.g., automatic teller machines, telephone menu systems, library search systems) or are designed for people who will use them over long periods of time and have the opportunity to learn the system (e.g., computer graphics, word processing, database programs). Interface design decisions should be based on the intended user group and the context of use.

Human–computer interface design issues have been studied for years, given that humans have been interacting with computers since the 1960s. Consequently, a starting point for designers is the general literature on human–computer interaction. Two valuable resources are the handbooks on this topic by Helander, Landauer, and Prabhu (1997) and Jacko and Sears (2003). In the Helander book, there are approximately 7,000 referenced articles, which illustrates the sheer magnitude of issues that need to be considered and the research that has been conducted in this field. How does a designer begin to consolidate this vast literature successfully to develop usable computer systems? Guidelines and best practices provide a valuable starting point. Generally accepted design principles are presented in Table 7.2. These general principles can serve as an initial starting point for system designers.

In the remainder of the chapter our focus will be on the specific needs of older users. As discussed in Chapter 1, chronological age is simply a marker for a constellation of experiences, capabilities, and

Table 7.2 Principles for optimizing human–computer interactions

Principle	Description	Examples
Compatibility	System design should be compatible wih user expectations	A knob turned clockwise results in an increase in something; counter-clockwise results in a decrease
Consistency	Location of items should be the same across screens; similar functions should act the same throughout the system	Save or home button should be in the same location on every screen; cancel button should always result in the action
Error recovery	Expect users to make errors and make recovery easy	Provide an "undo" option and meaningful error messages
Feedback	Results of actions should be clear	Provide status information such as an hourglass to indicate processing
Individualization	Enable the user to tailor the system to individual capabilities and preferences	Flexibility in display characteristics such as size of icons; more than one option to perform a task (e.g., menu versus control keys)
Memory	The user's memory should not be overloaded; memory aids should be provided	Do not require multiple meaningless steps to perform an action (CTRL-F-Q-L-R); provide labels to support memory
Structure	Provide structure to support performance	Sysem layout chart; site map; organized displays
Workload	Reduce information processing requirements of user	Organize displays and highlight critical information to reduce need for scanning

Sources: Sanders and McCormick (1993); Smith and Mosier (1986); Wickens (1992).

limitations. There are, however, a number of issues related to aging that are relevant to interface design and that need to be considered when designing systems that are likely to be used by older people.

Do older adults use computer technologies?

Relative to other age groups, the percentage of adults who are beyond age 50 and use computers is lower, but the usage rates continue to grow. The growth in usage rates from 1997 to 2001 is shown in Figure 7.1A for computers in general and in 7.1B for the Internet.

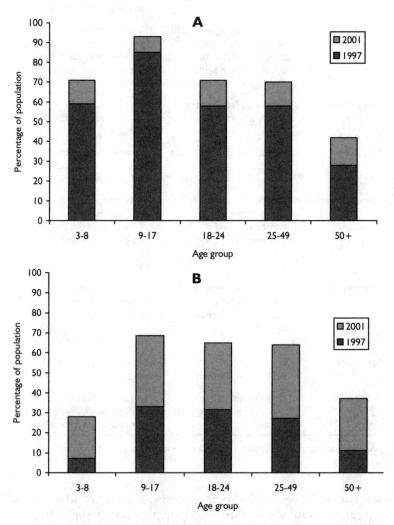

Figure 7.1   (A) Percentages of the population who used computers in 1997 and 2001 as a function of age group. (B) Companion data for use of the Internet. (Source: Kinsella and Velkoff, 2001).

These percentages represent many thousands of computer and Internet users of all ages. In addition to citing their interactions with computers, older adults report encounters with a broad range of technologies in their daily lives, including home security systems, multiple-line telephones, credit card scanners, video cassette recorders, computerized telephone menus, answering machines, entertainment centers, copy machines, cameras, microwave ovens, and fax machines. In today's society, it is becoming increasingly difficult to avoid interacting with computer systems. As a result, many older adults encounter interfaces in their daily activities whether by choice or by necessity.

Issues in aging and interface design

Multiple levels of user abilities

Cognitive abilities, such as working memory, attention, and spatial ability, are predictive of performance in using computer systems. Moreover, experience with technology in general is related to successful use of computer systems. As discussed in Chapter 1, older adults as a group tend to be more heterogeneous as compared with younger adults. That means that older adults vary a great deal in their capabilities, limitations, and experience with technology. Consequently, designers must ensure that systems that will be used by older adults will be usable by individuals with a range of abilities and experience.

One consequence of having less experience with computer technologies is that older individuals may lack the knowledge base required to interact with the system effectively. For example, older adults are less likely to know how to use windows or scrolling operations or how to make use of Boolean operators (i.e., "and" to narrow a search, "or" to broaden a search) when searching for information on the Web or online library systems. Older adults are less aware that these strategies are even available to assist them in effectively searching for information.

The information necessary for performing different tasks should be readily available rather than expecting the user to know intuitively how to perform these operations. Similarly, users should not be expected to remember sequences of actions (e.g., a key sequence); rather, the sequence should be available and visible in the interface.

Older adults tend to rely more on system tools, making it critical that such tools are well designed and unambiguous.

User goals and expectations

The degree to which users can successfully accomplish a task on a system is dependent on whether their goals match how the system functions. Additionally, their mental representation of the system must be veridical with respect to how the system actually works. To illustrate, when an individual uses an automatic teller machine, the selection of "fastcash" has certain consequences: there will be only a limited set of cash amounts that can be withdrawn, and only that single transaction will be allowed. These facts are not evident from the display; users must have this information in their heads or they will make the wrong selection. Even if a user has developed a mental model of how the system functions, it may not fit with the reality of how the system works. Or, as is the case with automatic teller machines, the systems may not be consistent across locations; for example, fastcash options at the airport are often larger ($50, $100, $200, $500) than those at the machine on a college campus ($10, $20, $50, $100). This variability in options across contexts may make it difficult for a user to develop an accurate model of the system.

Older adults' expectations about how systems should work may be based on how nonelectronic versions were structured. For example, many older adults have experience with paper-based card catalogs in the library, human tellers at the bank, manual cameras, and analog alarm clocks. These experiences have resulted in mental models that yield expectations about how an electronic or software system might work. A mismatch between the user's model and the designer's model is a prescription for failure. It is incumbent on designers to ensure that the system functions match user expectations. A less optimal solution to matching expectations to actual function is to provide training that enables users to develop the appropriate mental model (see Chapter 5). It is better to provide the knowledge "out in the world" (by proper design) than to create the knowledge "in a person's head" (by training). As much as possible, designers should use standard layouts across screens and applications to enable users to develop appropriate expectations about how systems function. Such consistency may be especially important for older adults, as their learning is more impaired by design inconsistencies.

Information organization

Interfaces contain information about how to accomplish a task or where to go, status information about what the system is doing, and help information. Designers thus have to make decisions about how best to present information, in what form, at what rate, and how much at one time. An understanding of the capabilities and limitations of older adults described in Chapters 2 and 4 can inform these decisions.

Processing speed slows down with age. As a result, older adults will have more difficulty with fast-paced speech, quickly scrolling text, or short-duration menu displays. Older adults require more time to process the information and more time to make a physical response (see Chapters 2 and 6).

Working memory capacity also declines as individuals grow older. Long lists of options presented on a telephone menu are likely to overload working memory. In fact, older adults use the repeat option more frequently than do younger adults to remember the content of lists presented on telephone menu systems.

Selective attention and memory for newly acquired information also show age-related decrements. These deficits combine to impact performance because older adults are more likely to forget command names and often must search through (often cluttered) displays for the relevant information. Even in comparatively simple displays, older adults require more time visually to search an array of information for a particular target piece of information. This difference is increased as the complexity of the display increases.

It is important to organize information to minimize working memory demands. Older adults benefit from the organization of information and from cues that reduce the search space. Consider searching computer displays in airport terminals for connecting flight information. Flight information at airports would be easier to locate if the user could constrain the search task by indicating a specific airline or destination city on information kiosks. Detailed flight information could also be provided on wireless networks that could interface with personal hand-held computers; such information could be downloaded and then sorted according to passengers' needs, thereby reducing visual search demands.

Information organization also influences the ability of older adults to use system interfaces easily. The options that are most important or most frequently used should be the most readily available.

Cluttered and complex visual displays should be avoided. Older adults take advantage of attentional cues, such as highlighting, and such cues should be used to support information search. Of course, older adults have to be able to see and hear the information being presented; these issues are discussed in Chapter 4.

Designers should also recognize that older adults have a wealth of knowledge that should be exploited in the system design process during the development of prototypes (see Chapter 3 for how to do this). Matching information organization in a display to the way older adults naturally organize the information has proved to be very successful. Moreover, information organization is often similar across adult age groups, so designing to support older adults will support younger adults as well.

Getting lost in the system and navigation tools

Relative to younger adults, older adults are more likely to get lost when navigating through complex systems, such as online library catalogs and the World Wide Web. Older adults revisit previously visited pages and screens more often and return to the starting point of the system to start a new task, even when that is unnecessary. These performance patterns may be due to age-related differences in working memory (i.e., they do not remember which screens they have visited) and to a lack of understanding about how the system works (e.g., not knowing that a new search may be initiated from any screen). Regardless of the reason, it is important for designers to recognize that older adults have navigation difficulties and to provide tools that can assist them.

Misunderstanding of system location can result in mode errors (see Chapter 3). For example, in using an online banking system, it is important to know whether the system is in transfer mode or history mode, as a particular keystroke sequence may initiate very different actions in each mode. If the system status is not clear, it is difficult to know what the consequences of an action might be. For example, an intention to click on an x to close a window may close an entire application instead. Incorrect awareness of the system mode could result in a deleted picture rather than a close-up view on a digital camera. Navigational support may reduce the frequency of mode errors, which may be more common for older adults owing to working memory and attentional deficits. Thus, navigational support not only involves helping users to move around, it involves assisting

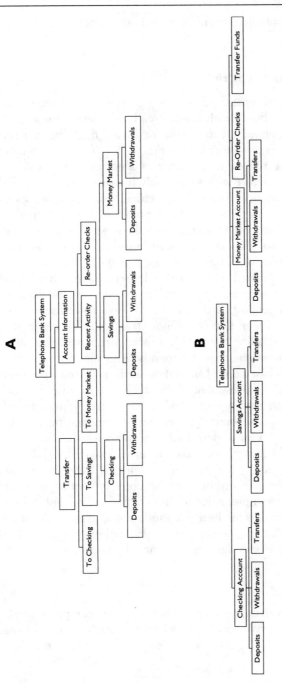

Figure 7.2 Hypothetical menus for a telephone banking system: (A) a deep menu structure; (B) a broad menu structure.

users in understanding where they are in the system at any point in time.

Depth versus breadth

During the development of any menu system, the designer must make decisions about depth versus breadth of menu structures. For example, a telephone menu system for a bank might be structured in different ways, as illustrated in Figure 7.2. There are tradeoffs between having a broad menu structure with only a few levels and having a deep structure with many levels. Important is that the trade-offs differ according to whether the menu is being presented visually or auditorily. For a visual menu, a broad structure reduces the demands on memory but also increases the need for visual scanning or scrolling through screens; for an auditory menu, the broad structure increases demands on working memory because a longer list of items is presented. Menus with a deeper structure may improve the organization of the information but also increase the likelihood that users will get lost in the system or have to backtrack to an earlier level, problems that occur more frequently for older adults.

What is the solution for the ideal menu structure? The perhaps unsatisfying answer is that it depends. The menu development process must be based on (1) an analysis of the tasks to be performed by the users; (2) an assessment of the labels to be used and whether they are familiar, unambiguous, and equally meaningful to different users; (3) a decision about the medium of presentation (visual or auditory); and (4) an understanding of whether the users of the system will be frequent users who may learn the system, or infrequent users. If older adults are part of the intended user population, designers should also consider whether some type of augmentation is feasible. For example, older adults benefit from a graphic display of the available options in a telephone menu system. As always, general principles may constrain the initial solution space, but user testing and iterative design are also critically important.

Compatibility

Compatibility is an important principle of human–computer interface design for users of all ages but may be particularly critical for older adult users. Consider a task in which information must be transferred from a paper form onto a computer system (e.g., a data entry task). Older adults are slowed to a greater degree when the display of the paper form is incompatible with the layout of the computer display (e.g., if the information on paper is displayed in columns but must be entered into rows on the computer display). Older adults benefit proportionally more than younger adults if the displays are made compatible.

Another example of incompatibility is a mismatch between options on a display and selection buttons. For example, some automatic teller machines have three options displayed on the right-hand side of the screen and four buttons to choose from. Similarly, credit card readers in grocery stores often have more buttons than options, making it unclear what the correspondence is between the options displayed and the proper button to be pressed. A related problem with credit card readers is the mapping between the options and the response keys: on the display, the "yes" option might be on the right with the "no" option on the left, but the response keys are in the opposite configuration.

Compatibility also relates to the correspondence between movements and whether something is increased or decreased. There are cultural constraints (i.e., population stereotypes) such that in North America, for example, up, right, and clockwise typically indicate "more," whereas down, left, and counter-clockwise typically indicate "less." This type of knowledge is part of a person's semantic knowledge base and is maintained into old age. As such, designers should take advantage of the constraints available by the conventions of the culture and exploit such constraints to improve the usability of systems.

The selection of labels for menus is also relevant to the issue of compatibility. A user's label for the task must correspond to the menu label for that task. The use of jargon or unfamiliar terms may be especially problematic for older adults because the need to decipher the terms and determine which one matches their goal adds extra demands to the task and may overload their working memory capabilities. However, the general organization of well-learned information is comparable across age groups and is well maintained into old age. Designers should capitalize on older adults' knowledge base to select the most compatible labels. The label selection process may reveal a natural (i.e., learned) organizational structure that would inform the depth-versus-breadth decision as well.

Documentation and error recovery

The documentation for a system includes error messages, user manuals, online help systems, and possibly videotapes or online tutorials. Unfortunately, this aspect of system design typically receives minimal attention during the design process. Designers assume that people do not read documentation. However, people do use documentation when they need it, if it is helpful. If the documentation is not informative, users will quickly learn to ignore it.

Poorly designed documentation is likely to have the biggest impact on the performance of older adult users. Relative to younger adults, older adults are more likely to want formal training to use a system, are more likely to request help while using the system, and are more likely to make errors and have to interpret error messages to figure out how to correct a problem.

It is almost inevitable that a user of a computer system will make an error at some point. Perhaps the error will be due to a misunderstanding about how the system works, or the error may be the consequence of a poorly designed system. In any event, errors do occur, and systems should be designed to be informative about what error has been made, what the consequences of the error are, and how to recover from the error. An example from a blood glucose monitor will illustrate a situation in which these guidelines were not followed. With this particular system, a user inserting a test strip upside down will receive the message "Error 2." Clearly, this message is not very informative. Similarly uninformative error messages are often presented on printers and on copy machines. Such messages provide no indication of what was done incorrectly, what the result of the action is, or how to correct the problem. This information should be provided by the system itself. If not, it should at least be in the user's manual (the manual for the blood glucose monitor explains that Error 2 means the system is not functioning properly—also not very informative).

Older adults tend to be more error-prone when using technologies. Given that they are also less frequent users of some types of technology, it is unlikely that they will be able to determine the source of an error on their own through their previous experience. Instead, they are going to be more dependent on the error messages provided by the system or explained in a manual. User testing of such feedback is crucial. Designers must understand the errors that might occur, the context in which the errors will occur, and what the user will need to know to be able to recover from the error. Perhaps not surprising, older adults have expressed some trepidation in using new technologies, owing in part to fear of making an error and not knowing how to correct it.

The information provided in the documentation should not only be clear and informative, it should be easily accessible. This is particularly relevant to the development of an online help system. Often, people need help when using advanced help systems! Searching for the proper terms to describe the problem is difficult for novice users who do not have a mental model that matches the system model (i.e., they are unable to formulate a description of the problem that matches one of the categories in the help system).

It is a fallacy that systems of any depth and breadth will be developed so that documentation is not necessary. The designer's mindset must be that documentation is an integral part of the system and that it includes such things as informative error messages and usable help systems (print or online). To ensure that documentation will support the system's use by older adults, this user group must be included in the development process of the documentation. An understanding of the errors they are likely to make, their knowledge base about the general domain and the specific system, and their preferences for how best to find assistance should be considered during the documentation design process.

Guidelines for effective interface design

To summarize the interface design guidelines that are important to consider in designing for an older population, we have grouped them into the following categories: physical characteristics, which may be influenced by age-related differences in perceptual and movement control capabilities; navigational issues related to maneuvering within the system; information organization; and more general conceptual issues.

Physical characteristics

*   Minimize clutter. Clutter can be visual (such as too many display items in any one location); auditory (too many sounds to make sense of such as warning tones); cognitive (such as too many things to keep in memory); and movement-related (such as too many response items that are too small).

- Allow adaptability (e.g., for Web pages, avoid style sheets that override the ability for users to increase font size).
- Establish appropriate temporal constraints for carrying out commands; ensure that drop-down or pop-up menu duration is long enough to be able to carry out the commands.
- Ensure that screen characters and targets are conspicuous and accessible (e.g., font size less than 12 point should be avoided); icons should be large enough to select easily; auditory information should be presented at the proper pitch, frequency, and rate (see Chapter 4 for more guidelines about physical characteristics).

Navigation

- Screen scrolling should be minimized, especially horizontal scrolling.
- Provide a site map.
- Provide search history: users need to know which pages have already been visited (e.g., in a list of items, change the color of pages previously visited).
- Indicate clearly where the user currently is, especially important if multiple windows or applications can be open or if different modes can be activated.
- Provide navigation assistance (e.g., how to link back to particular points in the system, not just going back to the home page but back to a previously relevant page).

Information organization

- Optimize information organization within natural or consistent groupings (e.g., group information that is related); keep most frequent operations highest in the menu structure.
- Develop the menu structure to match the medium of presentation; for visual displays, use broader menus with less depth; for auditory displays, use deeper menus with less breadth.
- Frequent and important actions should be easily visible and accessible (how to check out, how to exit from a system, how to save a file).

Conceptual

- Provide a standardized format within and, if possible, between applications (e.g., error messages should always appear in the same location); provide links to a home page or help in a consistent location.

- Ensure compatibility. Capitalize on user expectations and population stereotypes.
- Clearly convey current system status: make clear which window is open or which option is active; make clear the consequences of an action.
- Provide feedback about task completion, confirmation of activity, current state information: have I paid the bill, have I placed the order, have I sent the e-mail, did I delete this file, which window is currently open or active?
- Enable easy error correction; minimize opportunity for making the error (e.g., use of "are you sure?" prompts); provide opportunities for error recovery through clear information about type of error and options for correction.
- Provide system tools to support user goals, such as methods for broadening or narrowing a search or guidance regarding common "next-steps" in a process.
- Design adaptability and system flexibility of the system for different user levels; ensure that the system grows with the experience and skill level of users (e.g., provide multiple methods for command execution, such as menus, icons, short-cut control keys); allow experienced users to reconfigure the display; allow advanced functionality, such as cascading windows.

Using these guidelines for facilitating proper interface design

Older adults are more likely, relative to younger adults, to be affected if a design principle is violated. Younger adults may be able to deal with inconsistencies and incompatibilities, but older adults will be less able to compensate for poor design. The themes of the design recommendations are two-fold: capitalize on the knowledge and capabilities of the user group, and provide environmental support for the limitations of the user group.

Understanding the labels that users have for functions, the ways in which they organize information, their expectations about how systems work, and their experience with similar systems will all contribute to the development of systems that are usable by that population. Recognizing that instructional support is desired and necessary and designing such support to be informative and effective is very important.

Environmental support involves providing information, such as cues, reminders, or system tools, to support the intended action of the user. This notion is analogous to the idea of putting the needed information for a task in the world, rather than requiring the information to be in the head of the user. Although the provision of environmental support may improve the performance of all users, it may be particularly useful for older adults. Many of the design guidelines are instances of developing appropriate environmental supports: making target information conspicuous, providing navigational aids and system status information, organization of information, standardization and compatibility, feedback, instructional support, error information, and system tools to support user goals. The development of environmental supports should be based on understanding of the behaviors and processes of the users who should be supported (e.g., working memory, selective attention, spatial processing).

The general literature on human–computer interaction provides principles and guidelines to follow for interface design. The discussion in this chapter highlights those issues that are particularly relevant for systems that will likely be used by older adults. However, all this information provides only a starting point for the design process. The guidance can in essence make clear the solution space by suggesting prototype designs that will optimize the performance of older users. However, as with any design, success is going to be determined by the initial assessment of user needs, formative and summative evaluations of prototype systems, usability assessment, and iterative design. These processes are described in more detail in Chapter 3.

Recommended readings

Helander, M.G., Landauer, T.K., and Prabhu, P.V. *Handbook of Human–Computer Interaction*. 2nd edn. Amsterdam: North-Holland, 1997.

Jacko, J.A., and Sears, A. *The Human–Computer Interaction Handbook: Fundamentals, Evolving Technologies, and Emerging Applications*. Mahwah, NJ: Erlbaum, 2003.
See especially Czaja, S.J., and Lee, C.C. Designing computer systems for older adults, 413–27;
Mehlenbacher, B., Documentation: not yet implemented, but coming soon! 527–43.

Liu, Y. Software-user interface design. In G. Salvendy (ed.), *Handbook of Human Factors and Ergonomics*. 2nd edn. New York: Wiley, 1997:1,689–724.

Nielsen, J. *Designing Web Usability: The Practice of Simplicity*. Indianapolis: New Riders Publishing, 2000.

Norman, D.A. *The Psychology of Everyday Things*. New York: Harper Collins, 1988.

Rogers, W.A., and Badre, A. The Web user: older adults. In A. Badre, *Shaping Web Usability: Interaction Design in Context*. Boston: Addison-Wesley, 2002:91–108.

Schneiderman, B. *Designing the User Interface*. 3rd edn. Reading, MA: Addison-Wesley, 1998.

Part III
Exemplar applications

Chapter 8

Making the work environment age-friendly

The preceding chapters have presented general information and guidelines relevant to design for an aging population. We now turn to a discussion of the application of these principals and guidelines within the context of work and work environments. Recent demographic trends underscore the importance of designing work environments and work tasks to accommodate both younger and older workers. Because of a number of factors, such as the aging of the "baby boomers" and changes in retirement policies and benefits, the number of older workers will increase substantially over the next two decades. It is anticipated that by the year 2010, the number of U.S. workers who are beyond the age of 55 will be approximately 26 million, a 46 percent increase since 2000; by 2025, this number will increase to approximately 33 million. Important is that there will also be an increase in the number of workers older than 65. In contrast to previous decades, labor force participation rates are projected to increase for both older women and older men, although the increase will be slightly greater for older women. The percentage of older workers is expected to increase across most occupational categories, with the largest growth occurring in white-collar occupations, such as executives and managers; professional occupations; sales; and administrative support (Figure 8.1).

Clearly, most businesses and industries now need to develop strategies to accommodate an aging workforce. Making work environments "age-friendly" requires understanding of (1) the characteristics of older workers, (2) the potential implications of aging for work, and (3) the demands and skill requirements of current jobs. As discussed in Chapters 2 and 4, aging is associated with a number of changes in functional abilities, some of which have implications for work. The goals of this chapter are to summarize what

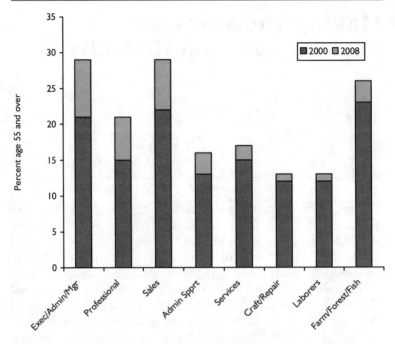

Figure 8.1   Percentage of workers aged 55 or older in different occupational groups, currently and projected for the near future. (Source: U.S. General Accounting Office, 2001.)

we know about aging and work performance, to dispel existing myths regarding older workers, and to provide some suggestions for designing work tasks and environments to maximize employment opportunities for and performance of older workers.

We place an emphasis on technology, as most workers interact with some form of technology in the performance of their job. Currently more than half (approximately 57 percent) of workers in the United States use a computer at work, and a fast-growing percentage of workers (currently approximately 42 percent) use the Internet or e-mail (or both) at work on a regular basis (Figure 8.2). Of course, use of computers at work varies substantially by occupation. The proportion of people using a computer at work is highest for those in managerial and professional occupations, followed by people in administrative support and technical occupations (e.g., computer programmers) and sales. As noted, the greatest growth of older workers is projected for managerial, professional,

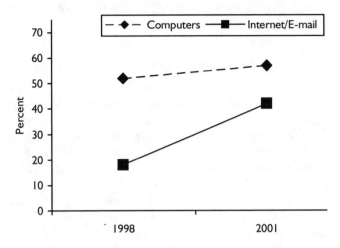

Figure 8.2 Percentage of individuals who use computers at work and the Internet or e-mail at work. (Source: U.S. Department of Commerce, 2002.)

administrative support, and sales occupations. Thus, older workers will have to interact with computers to meet job demands. The most common uses of computers at work include connecting to the Internet, e-mail, word processing, and working with spreadsheets and databases.

The prevalence of telecommuting is also rapidly increasing. In 1995, at least 3 million Americans were telecommuting, and this number is expected to increase by 20 percent per year. Telecommuting may promote employment of older workers, as telecommuting provides more opportunities for flexible work schedules and work at home. This may be especially beneficial to older adults who have problems in driving or have mobility limitations due to a chronic or functional impairment. However, telecommuting may limit opportunities for social interaction with friends and colleagues and thus exacerbate problems with social isolation. There is also some evidence that suggests that older people enjoy work that has a social component, especially those who are alone or isolated. Thus, it is important to consider the social implications of these types of tasks. Development of strategies to ensure that people have adequate training and technical support are also needed. These types of support services are not typically available within home environments. It is also important for employers to provide a means to keep workers informed of any changes in work-related information, job responsibilities, or organizational policies and procedures.

Technology and older workers

The influx of technology into the workplace has important implications for older workers. Technology influences employment opportunities and the types of jobs that are available. For example, such computer occupations as software engineers,computer support specialists and communication-network administrators will account for some of the fastest growing occupations over the next decade. In contrast, opportunities will decline within other occupations, such as office support personnel (e.g., typists) and bank tellers. Technology also changes the way in which jobs are performed. This means that workers need to upgrade their knowledge, skills, and abilities to avoid problems with obsolescence. Workers need to learn not only how to use technical systems but new ways of performing jobs. For example, customer service representatives typically handle customer queries by telephone and have some form of a social interaction with the customer. It is now possible to perform this type of task using a computer and e-mail. This implies that workers have to learn to use computer technology and e-mail and adapt to a new form of communication and customer interaction.

These are important issues for today's older workers, as they are less likely than younger workers to have had exposure to computer technology over their lifetime. As was illustrated in Figure 7.1, although the use of computers and the Internet is increasing among older adults, it is still low as compared to use by other age groups. In addition, because many employers perceive that older people are "techno-phobic" and unable to learn, older workers are often bypassed for retraining programs and are not given an opportunity to learn new skills involving technology use.

Advances in computing and telecommunications technology have made possible the rapid access to a vast amount of information and an array of resources in a wide variety of forms (e.g., text, graphics, speech). The Internet provides access to a vast number and variety of data sources on a wide variety of topics. Consider, for example, the ability of a manger or a financial planner to access information about a company or product prior to making an investment or marketing decision, or consider the ability of a health care specialist to gather the latest information on the management and treatment of a disease. Accessing and effectively using these databases requires navigating (often poorly designed) Web sites and filtering and integrating large amounts of information. Users are also required to have some knowledge of window operations, how to search for information, and how to use a mouse. These demands may be problematic for older adults because of age-related declines in abilities, such as working memory, reasoning and problem solving, and such abilities are predictive of performance using computer systems. Older adults may also lack the knowledge base (e.g., search strategies) to interact with these systems effectively.

However, as we illustrate throughout this chapter, many age-related declines in abilities can be compensated for through training and good design practices. For example, a recent study showed that older adults had more difficulty in performing a windows-based account-balancing task commonly performed within the banking industry. The task required participants to manipulate windows and integrate information from several databases to reconcile customer accounts. The results indicated that older adults had more difficulty in performing this task than did younger adults and that in addition to age, prior computer experience and such abilities as attention and working memory were important predictors of performance. Many of the older adults needed training on basic mouse and window operations prior to training on the account-balancing task. The data also indicated that design changes, such as providing on-screen aids, highlighting important information, and eliminating the need to "double-click" when using the mouse, resulted in performance improvements for all participants.

Unfortunately, designers of most systems have not considered older adults as active users of technology; thus, many interfaces are designed without accommodating the needs of this population. As discussed in Chapter 7, problems with usability, such as poorly designed displays or complex software applications, may make it difficult for older people to interact successfully with technology.

On the positive side, in many cases technology reduces the physical demands of work; thus, employment opportunities for older people may increase. As noted, computers also make work at home a more likely option. Adaptive technologies may also make continued work more viable for older people, especially those with some type of chronic condition or disability. For example, there are a number of technologies available, such as speech synthesizers, optical character recognition systems, or screen enlargement software, that can help people with low vision to function in the workplace. Similar technologies are available for people with other types of disabilities, such as hearing loss or mobility impairments. However, in order for these technologies to improve employment opportunities for older people, they must be available and useable, and people must be trained in their use.

We have conducted a number of studies examining the ability of older people to perform a range of technology-based work tasks, including data entry, inventory management, and a telecommuting customer service task. Overall, we have found that older adults are willing and able to perform these types of tasks although they may require more training and, on some measures of performance, such as those related to speed, they may not perform at the same level as that of younger people. However, we have also found that their performance improves with experience and that design interventions, such as changes in training protocols (e.g., providing training on basic computer operations), the design of display screens, or providing environmental support aids, such as a graphical overview of the system, typically improve the performance of people of all age groups. These types of interventions are discussed throughout this book. As discussed in Chapter 3, prior to implementing any type of design intervention, it is important to test them with representative groups of users.

Aging and work performance

What we know about older workers: myths and realities

Common myths about older people are that they are less productive, less able, less interested in work, and less willing than younger people to learn new skills. What are not myths are that the current generation of older adults is healthier, more diverse, and better educated than previous generations. They are also more interested in remaining engaged in some form of productive work. In fact, recent studies indicate that most older workers would prefer to continue to be engaged in some kind of work after their retirement and that a significant number of full-time retirees say they would like to be employed. The trend toward early retirement is also declining. Many middle-aged and older workers view the time approaching retirement as a time for midlife changes and are increasingly seeking more flexible work arrangements, such as working at home, reduced working hours, or part-time work. Many middle-aged and older people are also interested in starting second or third careers. Similar to younger workers, older workers also hold a wide variety of jobs; however, there is some variability according to age. Approximately the same percentages of middle-aged and older workers are employed in white-collar occupations, but older workers tend not to be employed in physically demanding or highly-paced blue-collar occupations.

Older workers are also willing and able to learn new tasks and skills. We found in our study of telecommuting, which involved people ranging from ages 50 to 80, that most of the older people were highly motivated to learn the job and interested in pursuing this type of work on the completion of the study. We have had similar findings in our studies of other computer-based tasks. Older people are also receptive to using new technologies if they perceive the technology as useful, the technology is easy to use, and they are provided with adequate training and support. They may experience more anxiety and less comfort than younger people when initially confronted with new technologies and tasks. However, we have found that if the technology is properly implemented and adequate training is provided, comfort tends to increase and anxiety tends to lessen as people gain experience interacting with these systems.

With respect to productivity, there is little support for the assertion that older people are less productive than younger people. The relationship between aging and work performance depends on the type of task, task experience, and training. For example, older people generally do not perform as well on tasks that are physically demanding or highly paced as compared to other types of jobs. In a study that examined age-related performance differences on simulated computer tasks that varied as a function of task complexity and pacing requirements, we found that older people did not perform as well as younger people on a data entry task that was "machine-paced." They were also more likely to report that the machine-paced task was stressful and fatiguing. However, for many types of tasks, job experience compensates for age-related performance differences. In our studies of age and computer-based work, we have found that prior experience with computers is a more important predictor of performance than is age. As noted earlier, it is especially important for older people to have adequate training in the actual use of the technology prior to learning a task, as they are less likely to have had exposure to technology, such as computers, than are younger people.

With respect to other measures of job behavior, the findings, while limited, are more conclusive. Regarding accidents, older workers tend to have lower accident rates than do younger workers; however, older workers tend to remain off the job longer if they are injured. Absenteeism and turnover rates also appear to be lower for older adults. Turnover rates are also lower among older adults; however, in many cases this is because of lower opportunities for alternative employment.

It is important to recognize that although there is a great deal of information about aging as a process, there are limited data on the practical implications of aging for work activities. Also, although we can discuss general trends regarding age-related changes in abilities, predictions about individuals' ability to learn a new skill or perform a particular job should be based on their functional capacity relative to the demands of that job rather than on chronological age. Also, across all occupations, there are wide differences in worker skills and abilities irrespective of age.

The potential implications of aging for work

Sensory and perceptual changes

There are a number of changes in abilities associated with normal aging that have implications for work. For example, as discussed in Chapter 4, there are a number of changes in vision that occur with age and, although most older adults do not experience severe visual impairments, they may experience declines in eyesight sufficient to make it difficult to perceive and comprehend visual information, such as written instructions, labels, or text on computer screens. This has vast implications for today's computer-oriented workplace, given that interaction with computer systems is primarily based on visually presented information. Visual decrements may make it more difficult for older people to perceive small icons on toolbars, read e-mail, or locate information on complex screens or Web sites. Age-related changes in vision also have implications for the design of written instructions and manuals and lighting requirements. As discussed in Chapter 4, the size, contrast, and spacing of text are important design considerations. Levels of illumination have to be higher for older adults, and potential sources of glare have to be minimized. Thus, it is important in designing workplaces to ensure that visual output screens (e.g., computer screens) are shielded from glare. Task lighting and low-vision aids may be particularly beneficial for older people. There are a variety of strategies available to accommodate people with low vision. These include strategies, such as providing larger monitors, increasing font size, or increasing screen resolution or use of speech as output device.

Many older adults also experience some decline in audition that has relevance to work settings. For example, older people may find it difficult to understand synthetic speech, as this type of speech is typically characterized by some degree of distortion. High-frequency alerting sounds, such as beeps or alarms on equipment, may also be difficult for older adults to detect. Changes in audition may also make it more difficult for older people to communicate in noisy work environments. Therefore, as discussed in Chapter 4, it is important to ensure that alerting or warning signals do not exceed 4,000 Hz. Ambient noise should also be minimized by using sound-absorbing materials on walls, floors, and ceilings and enclosing sources of noise where possible.

Psychomotor skills and strength and endurance

Aging is also associated with changes in motor skills, including slower response times, declines in ability to maintain continuous movements, disruptions in coordination, loss of flexibility, and greater variability in movement. Also as discussed in Chapter 9, the incidence of chronic conditions, such as arthritis, increases with age. These changes in motor skills may make it difficult for older people to perform tasks, such as assembly work that requires manipulation of small items, or to use current input devices, such as a mouse or keyboard. We have found that older people often have difficulty in performing mouse control tasks, such as pointing, double-clicking, or dragging. The timing of the mouse can also be changed to accommodate longer response times between clicks. Alternative input devices can also help to alleviate problems with mouse control. As discussed in Chapter 6, it is important to match the input device with user characteristics and the demands of the tasks. For example, a track ball is preferable to a mouse for novices if the operation involves double-clicking, and a light pen is preferable for pure pointing tasks.

Older adults also tend to have reduced strength and endurance.

There is, of course, a great deal of variability in muscle groups, in types of muscular performance, and between individuals. However, in general, older adults are less willing and able to perform physically demanding jobs. Several studies have shown that workers in physically demanding jobs retire earlier than do those in less physically demanding jobs and that older workers are more likely to transfer to jobs that are less physically demanding. Thus, such tasks as construction work, fire fighting, law enforcement, or aspects of manufacturing may prove difficult for workers in their 60s or 70s. However, these conclusions are speculative, and there are little data regarding specific jobs or occupations. As noted, it is important to match the demands and the requirements of a task with functional capabilities of persons.

The application of general ergonomic principles to the design of workplaces is also particularly important for older people. These include guidelines related to placement of controls and storage units, placement of computers, and workplace layout. The need for extensive bending, lifting, or carrying of objects should also be avoided.

Cognition and learning

Age-related changes in cognition (see Chapter 2) also have relevance to work activities, especially in tasks that involve the use of technology. Declines in working memory make it difficult for older people to learn new concepts or skills or to recall complex operational procedures. However, this does not imply that older people are unable to learn to perform new tasks or use new equipment. As discussed in Chapter 5, it may take them longer or they may require more training. However, it is important that training programs and instructional materials are designed to accommodate age-related changes in perceptual and cognitive abilities. A number of studies have shown that training interventions can be successful in terms of improving performance. For example, we found that an active goal-oriented training approach was successful in terms of teaching computer-naïve older adults to learn text editing. Technology may also be used as a training aid. For example, interactive online training programs might be useful in terms of updating workers on general work methods or procedures, as these programs facilitate self-paced learning. However, careful attention must be given to the design of these types of packages. Strategies must also be developed to ensure that older adults are provided with equal access to training programs.

Simple changes in the design of jobs and equipment or the provision of job aids can also diminish the working memory demands associated with many tasks and be beneficial for older workers. In our study of telecommuting that involved searching through several databases to respond to e-mail queries from customers, we used a split screen so that the content of the e-mail was always visible. This eliminated the need for workers to remember the questions and issues that they needed to address. In our study of a computer-based data entry task that involved detecting illogical routes, we eliminated the need to remember the geographical location of states by using a pop-up menu that contained this information.

Declines in attentional capacity may make it difficult for older persons to perform concurrent activities, such as driving while processing information from a global-positioning system, or to switch their attention between competing displays of information. They may also have problems in attending to or finding information on complex displays, such as overly crowded Web sites. Highly paced work may also be unsuitable for older workers. We found, for example, that older people tend to find highly paced tasks, such as paced data entry tasks, more stressful and fatiguing than would younger people. They also had lower work output than that of the younger adults. However, in controlling for differences in speed, there were no age differences in accuracy. We also found that increasing the compatibility between the information content and layout of information forms and data entry screens decreased keystroke errors among both younger and older people.

Conclusion

In general, a number of issues must be considered in designing work and work environments to accommodate an aging workforce. They include identifying specific components of jobs that are limiting for older adults, and targeting areas where workplace interventions could best be used to enhance the ability of older people to meet job requirements. This chapter highlights some of these issues and provides examples of potential design interventions. These interventions might include changes in job design, workplace and equipment redesign, or the development of new and innovative training strategies. Other design guidelines and practices that are discussed in other chapters of this book are relevant to this topic. Of course, the application of any of these guidelines or interventions strategies must be based on a human-factors engineering approach and involve user testing with representative user groups performing representative tasks. We also need to consider both the positive and negative implications of technology for an aging workforce. It is also important to apply principles of good design to workplace technologies so that workers of all age groups can use these technologies.

Recommended readings

Charness, N., and Bosman, E.A. Human factors and age. In F.I.M. Craik and T.A. Salthouse (eds), *The Handbook of Aging and Cognition*. Mahwah, NJ: Lawrence Erlbaum Associates, 1992:495–551.

Czaja, S.J. Technological change and the older worker. In J.E. Birren and K.W. Schaie (eds), *Handbook of the Psychology of Aging*. 5th edn. New York: Academic Press, 2001:547–68.

Kroemer, K., Kroemer, H., and Kroemer-Elbert, K. *Ergonomics: How to Design for Ease and Efficiency*. 2nd edn. Upper Saddle River, NJ: Prentice Hall, 2001.

Panek, P. The older worker. In A.D. Fisk and W.A. Rogers (eds), *Handbook of Human Factors and the Older Adult*. New York: Academic Press, 1997:363–94.

Salthouse, T.A., and Maurer, T.J. Aging, job performance, and career development. In J.E. Birren and K.W. Schaie (eds), *Handbook of the Psychology of Aging*. 4th edn. New York: Academic Press, 1996: 353–64.

Chapter 9

Maximizing the usefulness and usability of health care technologies

Individuals beyond age 65 are likely to suffer loss of visual acuity and some level of hearing loss as well as reduced cognitive capacities, such as memory and attention (see Chapters 2 and 4 for more details). In addition, older adults often have at least one chronic condition, such as arthritis, hypertension, or diabetes (Figure 9.1), and many have multiple chronic conditions. Many women and men older than 65 are also caregivers, caring for a spouse or an elderly parent.

Such activities as coordinating physician appointments, remembering to take medications, processing information about recently diagnosed ailments, monitoring chronic conditions, or learning new medical procedures are part of the normal daily routine of many older adults. Increasingly, these activities involve the use of some form of technology, such as interacting with an automated reminder system, using the World Wide Web to access medical information, or using an electronic medical device. Technology offers the potential to help older adults to access health information or information about available resources and be actively involved in the management of their health care. For example, there are Web sites that offer tips for family caregivers, modules on nutrition and exercise, and general information about various disease and health care issues.

Technology can also be used to facilitate health care delivery to older adults and their families. Many older people have difficulty in accessing routine care because of logistics problems, such as transportation. Telemedicine technologies now make "virtual" house calls a reality. However, for such technology to be effective and safe to use, attention to human-factors principles and the needs of the user population is critical.

The purpose of this chapter is to illustrate the importance of the issues discussed in the previous chapters to the domain of health

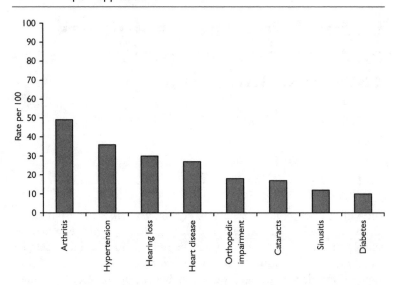

Figure 9.1   Most frequently occurring conditions per 100 older adults. (Source: http://www.aoa.gov/aoa/stats/profile.12.html.)

care. The focus of the discussion will be on technologies related to health care wherein users of the system are older adult patients or caregivers rather than health care practitioners. However, it is important to note that within the context of the work place, older physicians, nurses, therapists, and technicians are increasingly faced with the need to interact with complex technologies. Thus, the capabilities and limitations of older adults must also be considered in the design of technologies in health care facilities (see Chapter 8).

The discussion of health care technologies includes examples from two broad categories of technology: medical devices and communication technologies. Medical devices include such products as a blood pressure monitor, a digital thermometer, a blood glucose meter, an apnea monitor, a home defibrillator, an infusion pump, or a ventilator. These are home care devices that are used by laypersons without medical training and outside of the supervision of a professional (they may be used in the home, at work, in the car, in the yard, etc.). Communication technologies that support health care include the telephone and videophone, telemedicine systems, electronic monitoring, and the World Wide Web.

What makes health care technologies unique?

In many ways, the design of health care technologies is no different from the design of any technology to be used by older adults. Designers must consider the cognitive and perceptual capabilities and limitations of older adults (see Chapters 2 and 4), provide the optimal training (see Chapter 5), select the appropriate input and output devices (see Chapter 6), and structure the interface to ensure a usable system (see Chapter 7).

However, in other ways, the context in which health care technologies are used is unique; thus, so are the design considerations. Users may be in a situation of stress and perhaps may be highly emotional, having recently been diagnosed with an illness or suddenly been called on to care for an ailing spouse. The domain is often not well known, and replete with unfamiliar terms and phrases (e.g., infusion, systolic, benzodiazepine, diabetes mellitus). The consequence of making a usage error could mean prolonged illness, an additional trip to the hospital, or worse. The emotions, stress, and the disease itself might exacerbate existing cognitive impairments; medications and diseases can worsen cognitive, perceptual, and motor difficulties. Medical devices may be complex and sequential, with multiple steps and many opportunities for errors; devices often have to be calibrated to function properly. The instructions for use and maintenance are typically minimal and often are not developed for understanding by a layperson. Feedback and error information is often vague or confusing. Communication may become overly complex simply owing to the nature of the information being conveyed and the circumstances of the situation. Many older people, particularly women, live alone and may be using these technologies without support or guidance from anyone. Together, this confluence of factors represents the unique situation of using technology in the support of health care.

Movement control, perception, cognition, and interface design

As with any technology, successful use of health care technology will be dependent on the match between the demands of the system and the capabilities of the user. User problems with some commonly used health care devices illustrate the role of cognitive, perceptual, and movement control factors for usability (Table 9.1). Notice that

Table 9.1 Usability problems reported for common health care products

	Reported usability problems		
	Middle Aged (36–54)	Young-old (55–64)	Old (65–91)
Glucose monitoring device			
Users reporting problems (%)	27	38	42
Text comprehension (%)	17	10	22
Symbol comprehension (%)	0	20	11
Perceptual (%)	17	20	22
Memory (%)	67	40	44
Movement control (%)	0	10	11
Electric toothbrush			
Users reporting problems (%)	14	14	26
Text comprehension (%)	10	0	11
Symbol comprehension (%)	10	0	0
Perceptual (%)	50	40	22
Memory (%)	20	40	22
Movement control (%)	20	20	44
Thermometer			
Users reporting problems (%)	24	24	31
Text comprehension (%)	2	16	7
Symbol comprehension (%)	11	5	7
Perceptual (%)	51	32	60
Memory (%)	16	16	0
Movement control (%)	20	32	27

Source: Data taken from product usage survey conducted by Hancock, Fisk, and Rogers (2001).

Note:
Percentages within categories of problem types are for users reporting problems.

the percentage of users reporting difficulties increases for the older adults. Also, note that cognitive difficulties (text and symbol comprehension, memory) are more evident for the comparatively more complex glucose monitoring device, whereas perceptual and movement control difficulties are more commonly reported for an electric toothbrush and thermometer.

Designing to accommodate age-related changes in movement control, perception, and cognitive functions may be particularly important for medical devices because illness and medications can intensify their severity. Older adults often have restricted fine-motor capabilities owing to age-related slowing of the movement control system, and the precision of their movement is influenced by the presence of "noise" in their neural system. Such movement difficulties are worsened by arthritis or conditions that cause tremors, such as Parkinson's disease. Medical devices that have very small parts (e.g., hearing aids) or require fine movements (e.g., inserting a catheter) likely will be more difficult for older users, and their limitations should be considered during the design process. Designers must also recognize that chronic conditions are often progressive and that systems must accommodate the changing capabilities of the user over time.

Design issues related to changes in perception also arise in the development of medical devices and communication systems that support health care. For example, visual displays on medical devices provide critical information about the status and output of such devices, yet many of them are quite small and difficult to read (e.g., battery indicator on an oxygen tank, displays on an apnea monitor or a blood glucose meter). Videophones may be used to present health care instructions, yet little work has been done to ensure that the clarity of these systems is sufficient for older adults to extract the critical information successfully.

As discussed in Chapter 4, many older adults are faced with declines in vision, which may be exacerbated by a chronic condition, such as diabetes or glaucoma. Designers must be aware that displays will typically not be viewed by young eyes; they must follow the guidelines presented in Chapter 4 and conduct user testing studies (see Chapter 3), not just with older adults but with older adults who have the condition for which the medical device or communication system is designed. Movement control difficulties must also be considered during user testing. For example, a person with tremors associated with Parkinson's disease may have particular difficulties in using an input device to search the Web or to manipulate the parts of a medical device.

Similarly, declines in hearing may impede older adults' capability to detect alarms (e.g., the low-battery alarm on a portable respirator). Older adults have particular difficulty with high-frequency sounds, and an alarm that is not designed with consideration for age-related hearing changes might not be heard when needed. Automated speech messages must be designed to accommodate age-related changes in speech comprehension and general hearing losses. It is important to note that accommodation does not mean that the information should simply be presented slower and louder (see Chapter 4). Older adults rely more on contextual cues and the nonverbal aspects of speech, such as rhythm and tone, and these qualities must be retained for recorded messages and synthetic speech. The messages must use clear concise language and not overload the working memory capabilities of the listener. Important content information must be made explicit and not require the listener to make inferences. There is some evidence that older adults benefit even more than younger adults from the addition of a graphic representation of the information being presented in an auditory messaging system: such dual coding (presenting information both visually and auditorily) might be a way to maximize comprehension by older adults.

Working memory issues and other cognitive factors must also be considered during the development of medical devices and communication systems. For example, multistep sequences of activities that must be remembered and carried out for such systems to work are going to be problematic for older adults. In such situations, design must start with the recognition of who the users might be (suffering from one or more physical ailments, stressed, a novice, and without transferable knowledge).

Guidelines about information organization, design of menus, presentation of feedback, and error information should be followed (see Chapter 7). The number of steps required to perform a procedure or to activate a system should be minimized (see the minimize-steps principle of Chapter 6). Knowledge should be presented in the world (by using labels, population stereotypes, or other design features that afford proper use) rather than expecting it to be in the user's head (and thus dependent on memory). Systems should provide users with support to remember the proper sequencing of steps; for systems wherein procedures are sequential, errors may cascade, and the cost of a mistake may be very high.

A user manual or a help system may be referenced under a high stress situation; thus, it is imperative that relevant information be clear and easily accessible when needed. Similarly, if a system itself is providing feedback on the system state, the feedback must be designed to be useful and understandable to the user. For example, a blood pressure monitor currently available on the market uses a sound to indicate that the system is working properly. Unfortunately, this device uses the exact same sound to indicate that the system is not working properly. For users to differentiate the intended message of the sound, they must know the current system mode. This type of design is likely to cause a mode error (e.g., assuming the system is working properly when it is not).

Recognizing and addressing the interface design issues is the first step in the development of prototype systems. Usability assessments must then be performed in contexts that are representative of the situation wherein the device will be used, and the people tested must represent the actual end user group.

Training issues

How do individuals learn to use medical devices? Sometimes, a prescribing physician will have the nurse give a patient a brief tutorial about the device. Other times, especially for devices available over the counter at the pharmacy or the local discount store, training consists of a user manual or videotape provided by the manufacturer. Sometimes, users just start using the device and attempt to learn it through trial and error.

Regardless of the mechanism of training, most evidence suggests that currently available methods are insufficient to enable users to use the device properly, recognize when they have made an error and understand how to fix the problem, or adequately maintain and calibrate the device. Important is that such problems tend to be more extreme for older adults.

Training problems are multifaceted. The training may be provided under the stressful and emotional context of being newly diagnosed with an illness. Training provided by a health care professional may be presented too quickly, using jargon, with little practice by the patient, and without adequate explanation of the difficulties that may arise if the steps are not followed properly. When users are at home attempting to use a system, they may forget the details of the steps, have no idea about what to do if the system does not operate as expected, and have no immediate access to help. This point is well illustrated with the following excerpt:

> Obtaining and using home medical equipment carries a great deal of emotional stress, and that affects how we learn about the equipment and how we use and maintain it. Before Leonard was discharged from the hospital, we had a training session. I tried to listen carefully, but it was overwhelming—so many procedures to remember, so much terminology ... When I got home, I realized I hadn't absorbed half of what I'd been told ... Twice we have gone to the hospital emergency room only to discover that we could have taken care of the problem at home.

But we're not doctors or nurses, and we can't always tell whether
a problem is urgent.

(Smith, Mintz, and Caplan, 1996, p. 6.)

Training systems developed by manufacturers often do not follow
principles of instructional design and have not been user-tested with
users, either younger or older. The development of the training and
documentation often receives the least attention in the development
process. Although this may be annoying in learning to use a videocassette recorder, it may be life-threatening in learning to use an
infusion pump. In addition, some home health care devices were
not initially intended for use in the home; they were intended for
use in health care settings. Consequently, existing labels and instructions may be geared toward health care professionals and hence be
less than understandable by laypersons.

How can training be provided to ensure that users can use medical
devices safely and effectively? First, it is important to recognize that
devices that appear to be simple to use by designers and people with
domain knowledge may be complex for novice users. Misrepresenting
a medical device as very simple to use can cause a variety of problems.
Users may assume that no training is required and be overly confident
that they are using the device properly. Or, users may believe that
difficulties that they experience are due to their own inadequacies
rather than to the complexity of the system itself. As well, the
functionality of the device may be underused if the system is not
well understood.

Second, training must be designed to match the demands of the
task and the instructional medium. The decision to use audio, or
video, or animation, or text-based materials should be based on an
analysis of the task and on an analysis of the advantages and disadvantages of a particular medium. For a task that has spatial demands,
such as requiring a particular relative orientation of components
(e.g., using a blood glucose meter), video instructions are superior
to text or audio instructions. However, for a memory-demanding
task (e.g., loading a medication organizer), there is little benefit from
the video information compared to audio-only.

Third, training must be provided for both long-term retention
and immediate mastery. Even if patients can use a device at the
doctor's office or in the pharmacy, they may not be able to use the
device at home a week later, or even 24 hours later. Long-term
retention of information is superior when training is distributed over

time (rather than massed) and when practice continues past initial mastery (i.e., overlearning).

A related issue is the importance of refresher training. Experienced users lapse into error-prone behaviors, such as taking shortcuts or omitting required steps; refresher training should be provided periodically to ensure that the specific procedure is being appropriately followed. Such refresher training may be required after shorter intervals for older as compared to younger adults. Users might be encouraged to review the instructions periodically and to calibrate their skills and knowledge, much as they must calibrate the system periodically.

Maintenance, calibration, and updating of the technology

Personal computers are generally quite reliable and operate without need for calibration or system maintenance (notwithstanding the need for software updates). If (when) a computer in the home fails, taking it to a computer repair store and living without it for a couple of days is a viable option: this absence may be annoying and inconvenient, but it is typically not life-threatening. The same is not true of home health care technologies. Many of these technologies require periodic calibration to ensure that they are functioning at their peak accuracy level. Moreover, such devices require periodic cleaning and maintenance to ensure optimal performance. Additionally, when a system fails, it may be extremely difficult to find someone with the knowledge necessary to fix it, and living without the device for several days may not be a viable option. Users must also be able to recognize when a system is malfunctioning, either as part of the training provided or through clear and meaningful alarms or feedback from the system. How can these issues be addressed? Options for these different issues are discussed next.

Maintenance

Systems should be designed to be easily maintained. Designers must recognize that, especially for expensive devices, systems may be used for longer than recommended and that those systems will have to be maintained. As long as a system is functioning, it will be used, and the user has to be able to maintain it (and calibrate it) to ensure that it is functioning adequately. There is anecdotal evidence to suggest that health care technologies are shared among patients or "handed down" from one relative to another. Or, due to the expense of replacing a system, a patient may continue to use it far longer than might be recommended by the manufacturer. Users must be able to maintain a device easily, or it must be clear that unless the device is periodically inspected and maintained by a professional, the device is unsafe to use. One strategy may be to design a mechanism whereby, if the device is not maintained sufficiently well to operate safely, the device does not operate at all. Perhaps a better solution would be that when systems are not working properly, users should be locked out with an unambiguous error signal (display/sound) and instructions of what to do next (e.g., call a 1-800 number).

Calibration

Designers should develop systems that (1) do not need frequent calibration, (2) are self-calibrating, or (3) are easily calibrated. An analysis of a popular blood glucose meter revealed that it required three calibration procedures. First, each time a new box of test strips was used, a code on the meter had to be set to match the code provided on the test strip vial. Second, the meter had to be calibrated periodically by using a special check strip. Third, to ensure that the meter was providing correct glucose readings, the user was instructed to perform regular tests using a glucose control solution. If such a complicated calibration system is necessary (a total of 52 steps), users must receive the proper training about how and when to do the calibrations. However, a better solution would be to simplify the calibration procedures.

Users must also be able to recognize that the calibration has not been done correctly or recently enough. This type of information can be incorporated into the design of the system itself via a reminder system. For example, safe and effective system use should not be reliant on a user's remembering to calibrate the device on a weekly basis (or doing anything else that needs to be done periodically). Remembering to do something periodically is a type of time-based prospective memory, a task at which older adults are notoriously poor. Instead, the device itself should be programmed wherever possible to provide a meaningful reminder about what needs to be done and when. Such a reminder changes the prospective memory task into an event-based task (e.g., when I see this message, I do this activity, or when I hear this alarm, I press these buttons). Older adults are much better able to remember such event-based tasks (as compared to time-based tasks).

Another alternative is to incorporate internal diagnostics into the system to enable self-calibration and to take users out of the loop altogether. One potential drawback of automating the system calibration is that users may then not monitor the system status as closely and hence not notice that it has not been properly calibrated. Automation of systems increases the need for well-designed feedback and alarm systems.

Updating

Systems should be designed in such a way that they can be easily updated to accommodate changing needs of patients, to interface with technology updates, or to enable the addition of new information to the system. For example, a telemedicine support system that provides information about services available in the community for caregivers and patients must be easy to update, as this type of information may change periodically.

Safety issues

Home health care technologies may introduce hazards into the home, such as electrical hazards or the presence of flammable materials. As such, safety must be a primary concern throughout the design process. Ideally, products should be designed to eliminate all hazards. If it is not possible to design hazards out of a system entirely, a guard should be provided against a hazard, such as an automatic shut-off. If a hazard can neither be designed out nor prevented, an adequate warning system is crucial for safe use. Knowledge about hazards and an appropriate system of warnings must be incorporated into both the training and the documentation for the system.

Hazard awareness

Understanding that a hazard exists is crucial to being able to avoid it. The home environment itself may introduce hazards that would not occur if a device were being used in a health care setting. For example, using oxygen in a small kitchen with a gas oven creates an explosion hazard; having to plug electrical devices in living areas can create a tripping hazard; and plugging high-energy devices into old wiring can lead to a fire hazard. System designers must consider the context in which their devices will be used and attempt to design out, prevent, or warn about such hazards.

Older adults are generally aware of many hazards that exist in their home currently; however, the introduction of new devices and systems leads to new hazards about which they must be educated. There are at least two paths that can successfully provide hazard information. First are formal instructional materials, such as pamphlets, recorded messages, or videos. Second is the illustration of hazards through instructive vignettes (e.g., providing a story about a person using the device in the home to demonstrate the hazards that can occur). This latter approach may be particularly beneficial to ensure that the hazards are well remembered by older adults.

Warnings

Medical devices can be dangerous to use, and the hazard may not be open and obvious (e.g., the danger of static electricity in a room wherein oxygen is used). Conveying this type of information is precisely the point of a warning system: to alert the user of a hidden hazard, to provide information about the consequences of that hazard, and to provide guidance about how to avoid the hazard. Older adults do take warnings seriously, they attend to warnings on products, and they believe that warnings are important. It is thus incumbent on designers to ensure that warnings are provided and are comprehensible to the older adult users of the system.

A warning must be noticed and encoded. Thus, the perceptual and attentional characteristics of the warning must be designed with the capabilities and limitations of older adults in mind. A warning must be comprehended and understood. To enhance comprehension, symbols must be user-tested with older adults, and the symbols should be accompanied by explanatory text. Textual information should be explicitly presented, reducing the need for inferences. Metaphors and similes should be used to link the information with the semantic knowledge base of a user.

Ultimately, compliance is the goal for warning systems. There is no evidence to suggest that older adults are unwilling to comply: The important thing is to provide them with the knowledge they need to understand the hazard, the consequences, and the options for avoidance.

Credibility

System users must determine whether information they are provided is credible (i.e., whether it is believable and should be trusted). Such credibility decisions are based on multiple sources of information, such as the trustworthiness of the source, experience with the system, referrals from friends and family, and knowledge about the domain. In the context of health care and technology, we discuss the issue of credibility in two primary areas: credibility of a Web-based source of health information and the credibility of information provided by a medical device.

The World Wide Web

The Internet contains a tremendous amount of information on a wide array of topics, including health care issues. The sheer number of sites related to a particular topic is immense; if one enters the keywords "arthritis information" on one popular search engine, the yield is more than 800,000 hits. Determining which information is credible is a challenge for all users. Although there may be nothing unique about the Web in terms of differentiating useful information from bogus, the ease of accessibility of such information implies that a person may be exposed to larger amounts of information than if they searched for information in the library or read health care magazines.

Older adults do use the Web, and one of the main reasons they report for using it is to search for health-related information. Older adults may be susceptible to what is termed a *gullibility error* (believing something that is untrue), which tends to occur for people who are unfamiliar with a domain or who are desperate for a service. Such errors may be reduced by first ensuring that the overall site is usable by older adults through proper organization of information, use of familiar terms or labels, and explicitly stated information about to whom the information is relevant (see Chapter 7). It is also important to provide explicit cues to users about the credibility of the information being provided. One key determinant of credibility is the source of the information; provide sources for information to assist users in determining credibility (cf. the National Institutes of Health vs. Aunt Nancy's Home Remedies).

Medical devices

Medical devices typically provide some type of health status information to users (e.g., blood pressure, temperature, blood glucose level, blood oxygen saturation, or general status information). Patients or caregivers must use this information to make decisions about changes in medication or nutrition or whether a visit to the physician is warranted. Hence, the accuracy and credibility of the information provided is critically important.

System errors lead to a decrease in credibility. Consider a blood pressure monitor that is not reliable: It provides overly high readings on some occasions and overly low readings on other occasions. As a result, a user may not believe the readings or may not respond appropriately to deviations or warnings, instead blaming the unreliability of the system. If a patient's blood pressure is abnormally high or abnormally low, ignoring the information could be disastrous. The same issue arises with any type of medical device that provides information to a patient or a caregiver.

A designer must do two things. The obvious first step is to engineer the system to be as reliable as possible. The second (perhaps less obvious) step is to convey the proper information to a user, which means providing an accurate estimate of the reliability of the system. Users must be able to make an informed decision about the credibility of the information being provided by the system. The challenge for designers is to determine how best to convey what is likely to be complex information about probabilities and estimations (e.g., the readings are 85 percent reliable, reliability is influenced by the humidity in the room). It is vitally important that older adults be able to understand the information that is presented. As such, working memory limitations must be considered, the information must be made explicit to minimize the requirement for inferences, simple language (jargon-free) should be used and, to the extent possible, the information should be linked to general knowledge that an average layperson would be expected to have (see Chapter 2). In addition, as explained in Chapter 3, appropriate user testing must be conducted to ensure that users understand the information that is conveyed by the system.

Conclusion

The successful design of health care technologies requires attention to all of the issues discussed in the preceding chapters of this book.

Appropriate human-factors techniques of task and person analysis must be employed along with user testing and assessment. It is critical to follow the guidelines for maximizing perception and comprehension of information, providing training and instructional support materials, selecting input–output devices, and designing interfaces as a starting point to the development of safe and effective systems. The examples provided in this chapter illustrate the importance of good design in this domain.

However, optimal design of health care technologies requires the additional consideration of other issues, such as those discussed in this chapter. This class of technologies is defined by a combination of factors, such as the unfamiliarity of the domain, the stress and emotionality of the context, the high cost of errors, and the complexity of the systems themselves. These factors may make the design of technologies used to support health care more complicated than the design of technologies used in other domains.

Recommended readings

Institute of Medicine. *To Err Is Human: Building Safer Health Systems.* Washington, DC: National Academy Press, 1999.

Morrell, R.W. *Older Adults, Health Information, and the World Wide Web.* Mahwah, NJ: Lawrence Erlbaum Associates, 2002.

Park, D.C., Morrell, R.W., and Shifren, K. *Processing of Medical Information in Aging Patients: Cognitive and Human Factors Perspectives.* Mahwah, NJ: Lawrence Erlbaum Associates, 1999.

Rogers, W.A., and Fisk, A.D. *Human Factors Interventions for the Health Care of Older Adults.* Mahwah, NJ: Lawrence Erlbaum Associates, 2001.

Rogers, W.A., Mykityshyn, A.L., Campbell, R.H., and Fisk, A.D. Analysis of a "simple" medical device. *Ergonomics in Design* 9 (2001), 6–14.

Sainfort, F., Jacko, J.A., and Booske, B.C. Human-computer interaction in health care. In J.A. Jacko and A. Sears (eds), *The Human–Computer Interaction Handbook: Fundamentals, Evolving Technologies, and Emerging Applications.* Mahwah, NJ: Lawrence Erlbaum Associates, 2003:808–22.

Part IV

Conclusion

Chapter 10

Synthesis and comments

As is evident from the preceding chapters, it is possible to translate research on aging into guidelines for designing for older adults. In some cases, we were able to be quite specific; in other cases, the guidelines were, by necessity, more general in nature. The focus of this chapter is the general themes that recurred throughout the book.

Themes

Older adults do use new technologies

Older adults are active users of technology, and designers should think of them as a viable user group. Technology has the potential to enhance the lives of the growing older population, through technology to support health care or enable people to work at home; through the development of in-home technological supports that can delay entry into nursing homes; and through support for transportation, communication, and leisure activities. Older adults are willing users of technologies if (1) the benefits are clear to them, (2) they receive adequate instruction about how to use a system, and (3) the system itself is easy to use.

Experience (or lack thereof) influences performance

Today's cohort of older adults has less experience in using computer technology; as a result, they are less able to transfer skills and knowledge about how one computer-based system works to learning a new computer-based system. Older adults' mental models of how products should function may differ from younger adults', not because they are older but because of their differential experience. At the same time, older adults have a well-developed semantic memory base (i.e., their world knowledge) that may guide them in their interactions with technology. Designers must take advantage of this knowledge base and develop products and systems that are consistent with representations, expectations, and experiences.

If it cannot be seen, heard, or manipulated, it cannot be used

Fairly universal changes that occur with age are changes in visual, auditory, and movement control capabilities. Age-related changes in some aspects of vision begin in early middle age (30–40). Auditory changes are more evident at later ages but are quite common across older individuals, especially men. Movement control slows and becomes more variable as individuals age. These changes are well defined and well documented. If something that is being designed is going to be used by older adults, those older adults must be able to see the features and components of the system, be able to hear any auditory information that is presented, and be able to manipulate the component parts in time and with the precision necessary to operate the system appropriately. Although this may appear to be obvious information, there are myriad products in our environments that were not designed with older adult users in mind. Perhaps it is not so obvious because designers need to have specific guidance about age-related changes in these abilities to be able to design products appropriately. We have tried to provide such guidance in this book.

Age-related declines in cognition influences performance

Although there is substantial variability across individuals in patterns of age-related cognitive changes, there are certain abilities that do tend to decline as people reach their 60s and beyond. One of the most common and pervasive changes is a decline in working memory, the ability to keep information active and available for processing. Working memory is an important component of many activities; hence, declines in working memory lead to a wide range of performance difficulties. There are many aspects of system interaction that rely on working memory, not all of which are immediately obvious: the ability to translate movement of an input device and related movement on a system display; organization of information on a visual display; navigation through a complex system; recollection of a list of options on an auditory display; integration of training materials; memory for sequential steps of a procedure (and the list could continue). If a system is designed without considering the working memory capacity of its users, it will not be usable.

Other cognitive factors, such as attention and spatial cognition, tend to show age-related deficits, and these abilities influence system performance. Designers should recognize these patterns of age-related changes in cognition, develop systems that compensate for declining abilities, and make use of those cognitive abilities that do not show age-related declines (e.g., semantic memory).

Environmental support, knowledge in the world, affordances

How can cognitive declines be supported in the design of systems and products? A recurring theme in this book was the idea of providing environmental support for the tasks that users have to perform. If a sequence of procedures must be performed, do not require a person to perform that sequence on the basis of memory of the procedures. Rather, it is important to have the information available and accessible about the sequence of actions and to provide cues for actions at the appropriate times. If it is important to be able to retrace one's steps, provide information about where users have been and what activities they have performed, rather than forcing them to rely on their memory. The idea is to provide the needed knowledge in the world, when and where the user needs the information, rather than requiring users to have the necessary knowledge in their heads.

A related idea is to design for affordances; wherever possible, structure tasks and devices so that the system itself affords the right action. Buttons afford pressing, knobs afford grasping, text affords reading, and tables afford organization of information in rows and columns. Affordances may be physical (such as buttons and knobs), perceptual (such as relying on principles of perceptual grouping), and cognitive (such as processing information in a left-to-right order for native English speakers).

Preferences do not necessarily map to performance

One design strategy is to design a system to be adaptable by users: to let users select the settings and functions of interest. However, preferences do not necessarily map to performance. Making a system adjustable is not going to be a panacea for design problems because if people do not know what they need, they will not adjust the options and functions correctly. Moreover, making systems that must be adjusted and calibrated by users may increase the complexity of the system by adding to the list of things a user has to do.

Training, documentation, and user help are necessary (and should be well designed)

Another recurrent theme is the need for training, training that is based on analysis of the tasks that users will have to perform and on analysis of the trainee population. There are instructional design principles that should be followed, and there is evidence that older adults may have unique training needs, relative to younger adults. In addition, system documentation and user help systems must receive more attention in the development process, must be designed with the user group in mind, and must be be user-tested.

User testing is crucial for successful design

Perhaps the most important thing for designers to remember is to involve users in the design process, early and often. We need to understand at the very start of the design process what the needs, expectations, and preferences of users are. As early as is feasible, users should be involved in testing prototypes and providing input about usability and system functionality. User testing should continue even after a product has gone to market, to understand the user problems that might arise over time, how patterns of use change, and how the next iterations of the product should accommodate user needs. It is simply a fact that designers cannot, by themselves, identify in advance everything that users will do: how people will use or even misuse a product, the misunderstandings that may arise, the confusions, and the expectations that users have about what systems should do. If older adults are likely to be part of a user group, they must be part of the user test groups.

Good design for older adults is good design for everyone

In most instances, systems that are designed to be easy to use by older adults will also be easy to use by other user groups. In fact, older adults can help designers to identify quickly usability problems that might arise for other individuals: people of all ages with vision impairments or movement control difficulties; people who will be using the system while doing other tasks simultaneously (which would be occupying part of their working memory capacity); and people with developmental disabilities that impair their cognitive function. These groups obviously are not equivalent to older adults, but good design for older users is typically going to be good design for everyone.

Development of future technology

New is always new

There will always be new technology, and there will always be age-related changes in movement control, perceptual capabilities, and cognitive functions (medical breakthroughs notwithstanding). Our examples were based on current technologies, but our goal was to provide principles and guidelines that will be relevant to future technologies; we believe that the design process will benefit from an understanding of the basic fundamentals of age-related changes and from consideration of them at the start of the design process. Even future older adult cohorts will be faced with learning to use new technologies because it is not possible to foresee what the future will bring in terms of technology changes in the next few decades. Experience with today's technology may show minimal transfer to use of tomorrow's technology.

New changes are not necessarily better

There are two themes in current technology design that may make systems less usable by older adults. First is the trend toward miniaturization, where everything is getting smaller, in part because the technology has developed to allow more power in smaller spaces. Tiny cell phones or personal digital assistants may be fine for the nimble fingers and strong eyes of an adolescent or younger adult, but this trend creates serious usability problems for older eyes and arthritic hands.

The second trend is sometimes called *function creep*, which is the development of multifunction systems (e.g., a cell phone that is also a personal digital assistant; a television that is also a computer). The reality is that most users, not just older adults, use only a limited portion of the functions available on a particular device or system. More consideration should be given in the design process to the development of systems that have only the functions that users really want and need, without the inclusion of seemingly superfluous functions that increase the complexity of the device.

Conclusion

What is the best way to design systems, products, and environments for older adults? First, designers must understand the user population (i.e., follow the maxim of *know thy user*). We do not mean to imply in any way that all older adults are the same. In fact, we emphasize the reality: older adults do not represent a homogeneous group. There are differences in rates of change, patterns of changes, life experiences, compensatory strategies, motivation, attitudes, and more. However, there are general age-related changes that tend to occur, and designers who understand these general patterns will develop systems that are more easily usable by older adults and probably by other user populations as well.

Second, the tools and techniques of human factors should be used to develop prototypes that can then be tested with representative users, doing representative tasks, in representative contexts. The design guidelines in this book provide a starting point for the development of prototypes. Rather than starting with a few points of references, designers should be able to focus on a solution space for features to allow an initial approximation of an appropriate design: how to display the information, how users will optimally be able to interact with the system, how instructional support and user help should be designed, and how interface elements should be structured. Using these guidelines to develop initial prototypes should yield more usable systems, but appropriate user testing is invaluable and very necessary for successful design.

In sum, successful design depends on the match between the capabilities of users and the demands imposed by a system (Figure 10.1). To the extent that designers understand the capabilities (and

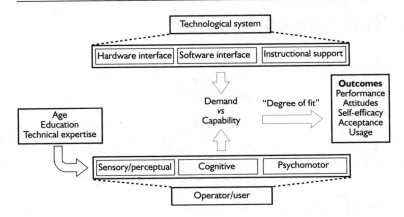

Figure 10.1 The CREATE perspective on designing for older adults. (Source: Adapted from Czaja, Sharit, Charness, Fisk, and Rogers, 2001.)

limitations) of older adult users, they will be better able to design systems that are usable by older adults.

Recommended readings

Czaja, S.J., Sharit, J., Charness, N., Fisk, A.D., and Rogers, W.A. The Center for Research and Education on Aging and Technology Enhancement (CREATE): a program to enhance technology for older adults. *Gerontechnology* 1 (2001), 50–9.

Fisk, A.D. Human factors and the older adult. *Ergonomics in Design* 7, 1 (1999), 8–13.

Fisk, A.D., and Rogers, W.A. *Handbook of Human Factors and the Older Adult.* Orlando: Academic Press, 1997.

Rogers, W.A., and Fisk, A.D. Human factors, applied cognition, and aging. In F.I.M. Craik and T.A. Salthouse (eds), *The Handbook of Aging and Cognition.* 2nd edn. Mahwah, NJ: Erlbaum, 2000:559–91.

References

Birren, J.E., and Schaie, K.W. *Handbook of the Psychology of Aging.* 5th edn. San Diego: Academic Press, 2001.

Brock, J.F. Computer-based instruction. In G. Salvendy (ed.), *Handbook of Human Factors and Ergonomics.* 2nd edn. New York: John Wiley, 1997:578–93.

BSR/HFES100 Human Factors Engineering of Computer Workstations. Santa Monica: Human Factors and Ergonomics Society, 2002.

Charness, N., and Bosman, E.A. Human factors and age. In F.I.M. Craik and T.A. Salthouse (eds), *The Handbook of Aging and Cognition.* Hillsdale, NJ: Erlbaum, 1992:495–551.

Charness, N., and Dijkstra, K. Age, luminance, and print legibility in homes, offices, and public places. *Human Factors* 41, 2 (1999), 173–93.

Craik, F.I.M., and Salthouse, T.A. *The Handbook of Aging and Cognition.* 2nd edn. Mahwah, NJ: Erlbaum, 2000.

Czaja, S.J. Aging and the acquisition of computer skills. In W.A. Rogers, A.D. Fisk, and N. Walker (eds), *Aging and Skilled Performance: Advances in Theory and Applications.* Mahwah, NJ: Lawrence Erlbaum, 1996:201–20).

Czaja, S.J. Technological change and the older worker. In J.E. Birren and K.W. Schaie (eds), *Handbook of the Psychology of Aging.* 5th edn. New York: Academic Press, 2001:547–65.

Czaja, S.J., and Lee, C.C. Designing computer systems for older adults. In J.A. Jacko and A. Sears (eds), *The Human–Computer Interaction Handbook: Fundamentals, Evolving Technologies, and Emerging Applications.* Mahwah, NJ: Erlbaum, 2003:413–27.

Czaja, S.J., Sharit, J., Charness, N., Fisk, A.D., and Rogers, W.A. The Center for Research and Education on Aging and Technology Enhancement (CREATE): a program to enhance technology for older adults. *Gerontechnology* 1 (2001), 50–9.

Fisk, A.D. Human factors and the older adult. *Ergonomics in Design* 7, 1 (1999), 8–13.

Fisk, A.D., and Rogers, W.A. *Handbook of Human Factors and the Older Adult*. Orlando: Academic Press, 1997.

Fozard, J., and Gordon-Salant, S. Changes in vision and hearing with aging. In J.E. Birren and K.W. Schaie (eds), *Handbook of the Psychology of Aging*. 5th edn. San Diego: Academic Press, 2001:241–66.

Gagné, R., Briggs, L.J., and Wager, W.W. *Principles of Instructional Design*. New York: Holt, Rinehart, and Winston, 1989.

Hancock, H.E., Fisk, A.D., and Rogers, W.A. Everyday products: easy to use ... or not? *Ergonomics in Design 9* (2001), 12–18.

Helander, M.G., Landauer, T.K., and Prabhu, P.V. *Handbook of Human–Computer Interaction*. 2nd edn. Amsterdam: North-Holland, 1997.

Institute of Medicine. *To Err is Human: Building Safer Health Systems*. Washington, DC: National Academy Press, 1999.

Jacko, J.A., and Sears, A. *The Human–Computer Interaction Handbook: Fundamentals, Evolving Technologies, and Emerging Applications*. Mahwah, NJ: Erlbaum, 2003.

Kinsella, K, and Velkoff, V.A. *An Aging World: 2001 (U.S. Census Bureau, Series P95/01-1)*. Washington, DC: U.S. Government Printing Office, 2001.

Kirwan, B., and Ainsworth, L.K. *A Guide to Task Analysis*. London: Taylor & Francis, 1992.

Kline, D. W., and Fuchs, P. The visibility of symbolic highway signs can be increased among drivers of all ages. *Human Factors 35* (1993), 25–34.

Kroemer, K., Kroemer, H., and Kroemer-Elbert, K. *Ergonomics: How to Design for Ease And Efficiency*. 2nd edn. Upper Saddle River, NJ: Prentice Hall, 2001.

Legge, G.E., Rubin, G.S., Pelli, D.G., and Schleske, M.M. Psychophysics of reading: II. Low vision. *Vision Research 25* (1985), 253–66.

Liu, Y. Software-user interface design. In G. Salvendy (ed.), *Handbook of Human Factors and Ergonomics*. 2nd edn. New York: Wiley, 1997:1,689–724.

Mehlenbacher, B. Documentation: not yet implemented, but coming soon! In J.A. Jacko and A. Sears (eds), *The Human–Computer Interaction Handbook: Fundamentals, Evolving Technologies, and Emerging Applications*. Mahwah, NJ: Erlbaum, 2003:527–43.

Morrell, R.W. *Older Adults, Health Information, and the World Wide Web*. Mahwah, NJ: Lawrence Erlbaum Associates, 2002.

Nielsen, J. *Designing Web Usability: The Practice of Simplicity*. Indianapolis: New Riders Publishing, 2000.

Nielson, J. *Usability Engineering*. Cambridge, MA: Academic Press, 1993.

Norman, D.A. *The Psychology of Everyday Things*. New York: Harper Collins, 1988.

Panek, P. The older worker. In A.D. Fisk and W.A. Rogers (eds), *Handbook of Human Factors and the Older Adult*. New York: Academic Press, 1997:363–94.

Park, D.C., and Schwartz, N. *Cognitive Aging: A Primer*. Philadelphia: Psychological Press, 2000.

Park, D.C., Morrell, R.W., and Shifren, K. *Processing of Medical Information in Aging Patients: Cognitive and Human Factors Perspectives*. Mahwah, NJ: Lawrence Erlbaum Associates, 1999.

Rogers, W.A., and Badre, A.The Web user: older adults. In A. Badre, *Shaping Web Usability: Interaction Design in Context*. Boston: Addison-Wesley, 2002:91–108.

Rogers, W.A., Campbell, R.H., and Pak, R. A systems approach for training older adults to use technology. In N. Charness, D.C. Park, and B.A. Sabel (eds), *Communication, Technology and Aging*. New York: Springer Publishing Company, 2001:187–208.

Rogers, W.A., and Fisk, A.D. Human factors, applied cognition, and aging. In F.I.M. Craik and T.A. Salthouse (eds), *The Handbook of Aging and Cognition*. 2nd edn. Mahwah, NJ: Erlbaum, 2000:559–91.

Rogers, W.A., and Fisk, A.D. *Human Factors Interventions for the Health Care of Older Adults*. Mahwah, NJ: Lawrence Erlbaum Associates, 2001.

Rogers, W.A., Meyer, B., Walker, N., and Fisk, A.D. Functional limitations to daily living tasks in the aged: a focus group analysis. *Human Factors* 40 (1998), 111–25.

Rogers, W.A., Mykityshyn, A.L., Campbell, R.H., and Fisk, A.D. Analysis of a "simple" medical device. *Ergonomics in Design* 9 (2001), 6–14.

Rowe, J.W., and Kahn, R.L. *Successful Aging*. New York: Pantheon, 1998.

Rubin, J. *Handbook of Usability Testing: How to Plan, Design, and Conduct Effective Tests*. New York: Wiley, 1994.

Sainfort, F., Jacko, J.A., and Booske, B.C. Human–computer interaction in health care. In J.A. Jacko and A. Sears (eds), *The Human–Computer Interaction Handbook: Fundamentals, Evolving Technologies, and Emerging Applications*. Mahwah, NJ: Lawrence Erlbaum Associates, 2003:808–22.

Salvendy, G. *Handbook of Human Factors and Ergonomics*. 2nd edn. New York: John Wiley and Sons, 1997.

Sanders, M.S., and McCormick, E.J. *Human Factors in Engineering and Design*. New York: McGraw-Hill, 1993.

Schneider, B., and Pichora-Fuller, M.K. Implications of sensory deficits for cognitive aging. In F.I.M. Craik and T. Salthouse (eds), *The Handbook of Aging and Cognition*. 2nd edn. Mahwah, NJ: Erlbaum, 2000: 155–219.

Schneiderman, B. *Designing the User Interface*. 3rd edn. Reading, MA: Addison-Wesley, 1998.

Smith, C., Mintz, S., and Caplan, A. Caregivers. In R.L. Klatzky, N. Kober, and A. Mavor (eds), *Safe, Comfortable, Attractive, and Easy To Use: Improving the Usability of Home Medical Devices*. Washington, DC: National Academy Press, 1996:5–7.

Smith, S., and Mosier, J. *Guidelines for Designing User Interface Software* (Technical Report NTIS No. A177 198). Hanscomb AFB, MA: USAF Electronic Systems Division, 1986.

Stanton, N. *Human Factors in Consumer Products*. London: Taylor & Francis, 1998.

Steenbekkers, L.P.A., and van Beijsterveldt, C.E.M. (eds), *Design-Relevant Characteristics of Aging Users*. Delft, The Netherlands: Delft University Press, 1998.

Swezey, R.W., and Llaneras, R.E. Models in training and instruction. In G. Salvendy (ed.), *Handbook of Human Factors and Ergonomics*. 2nd edn. New York: John Wiley, 1997:514–77.

Tinker, M.A. *Legibility of Print*. Ames, IA: Iowa State University Press, 1963.

United States Department of Commerce. *A Nation Online: How Americans Are Expanding Their Use of the Internet*. Washington, DC: U.S. Department of Commerce, 2002.

United States General Accounting Office. *Older Workers: Demographic Trends Pose Challenges for Employers and Workers*. Washington, DC: U.S. General Accounting Office, 2001.

Whitley, B.E. *Principles of Research in Behavioral Science*. 2nd edn. Boston: McGraw Hill, 2002.

Wickens, C.D. *Engineering Psychology and Human Performance*. New York: Harper Collins, 1992.

Wickens, C.D., Gordon, S.E., and Liu, Y. *An Introduction to Human Factors Engineering*. New York: Longman, 1998.

Index